APlusPhysics: Your Guide to

Regents Physics Essentials

Dan Fullerton

Physics Teacher
Irondequoit High School

Adjunct Professor
Microelectronic Engineering
Rochester Institute of Technology

Dedication

To AF and Piglet, you make my heart smile.

Credits

Thanks To:

my family and friends for their ongoing support and encouragement;
my students, because teaching physics is the best job in the world due to you;
my colleagues, Mike, Tom, Art, Laurie, Kevin, and Karen for their advice and candor;
Karl, Doni, Wade, and Patti, for teaching me to learn by allowing me to make mistakes;
Andrew, for proving anything is possible with hard work and determination;
Ed, for otherworldly patience in teaching me how to write;
Mike and Santosh, for modeling how to teach;
and Mr. Genung, who set the foundation for everything else.

Silly Beagle Productions
656 Maris Run
Webster, NY 14580
Internet: www.SillyBeagle.com
E-Mail: info@SillyBeagle.com

Cover Design: Dan Fullerton
Interior Illustrations by Dan Fullerton, Jupiterimages and NASA
All images and illustrations ©2011 Jupiterimages Corporation and Dan Fullerton
Edited by Joseph Kunz

Sales and Ordering Information
http://www.aplusphysics.com/regents
Sales@SillyBeagle.com
(585) 943-9449
Volume discounts available

Printed in the United States of America
ISBN: 978-0-9835633-0-3

Silly Beagle Productions

Welcome to APlusPhysics: Your Guide to Regents Physics Essentials. From mechanics, electricity and magnetism to waves, optics, and selected modern physics concepts, this book is your essential physics resource for use as a standalone guide; companion to classroom texts; or as a review book for your NYS Board of Regents Physics Examination.

What sets this book apart from the other review books?

1. This book is designed to provide everything you need to know to score well on the NY Regents Physics Exam.
2. It includes more than 500 sample questions with full solutions, all of which are Regents Exam or Regents-Exam-like questions.
3. Two Regents Physics Exams included, with solutions!
4. This book is supplemented by the free APlusPhysics.com web site, which includes:
   a. Animations and videos of selected physics concepts
   b. Interactive practice quizzes based on Regents Exam questions
   c. Podcast reviews for iPod, iTouch, iPhone, iPad, or any iTunes-capable device
   d. Integrated discussion and homework help forums supported by the author and fellow readers
   e. Student blogs to share challenges, successes, hints and tricks
   f. Latest and greatest physics news
   g. Thousands of practice problems

Just remember, physics is fun! It's an exciting course, and with a little preparation and this book, you can transform your quest for essential physics comprehension from a stressful chore into an enjoyable and, yes, FUN, opportunity for success.

How to Use This Book

This book is arranged by topic, with sample problems and solutions integrated right in the text. Actively explore each chapter. Cover up those in-text solutions with an index card, get out a pencil, and try to solve the sample problems yourself... before looking at the answer. If you're stuck, don't stress! Post your problem on the APlusPhysics website (http://aplusphysics.com/regents) and get help from other students, teachers, and subject matter experts (including the author of this book!) Once you feel confident with the subject matter, take one of the practice tests in the appendix and see how you performed. Review areas of difficulty, then try again and watch your scores improve!

Table of Contents

Chapter 1: Introduction

When asked what the most important factor was in developing The Theory of Relativity, Albert Einstein replied:

"Figuring out how to think about the problem."

What Is Physics?

So, you've decided to take on the challenge of learning physics and are now wondering what you have gotten yourself into. What do you hope to accomplish with this newfound knowledge? Where ever will you use what you learn? Why go to all this trouble?

Those are some pretty deep questions, which hopefully you've started to think about if you haven't previously. The answers, though personal, may be helped by first answering the question, "What is Physics?"

Physics is many things to many different people. If you look up physics in the dictionary, you'll probably learn physics has to do with matter, energy, and their interactions. But what does that really mean? What is matter? What is energy? How do they interact? And most importantly, why do we care?

Physics, in many ways, is the answer to the favorite question of most 2-year-olds: "Why?" What comes after the why really doesn't matter. If it's a "why" question, chances are it's answered by physics. Why is the sky blue? Why does the wind blow? Why do we fall down? Why does my teacher smell funny? Why do airplanes fly? Why do the stars shine? Why do I have to eat my vegetables? The answer to all these questions, and many more, ultimately reside in the realm of physics.

Matter

If physics is the study of matter, then we probably ought to define matter. **Matter**, in scientific terms, is anything that has mass and takes up space. So what's mass? **Mass** is, in simple terms, the amount of "stuff" an object is made up of. Phrased more practically, matter is anything you can touch – from objects smaller than electrons to stars hundreds of times larger than our sun. From this perspective, physics is the mother of all science. Astronomy to zoology, all other branches of science are subsets of physics, or specializations inside the larger discipline of physics.

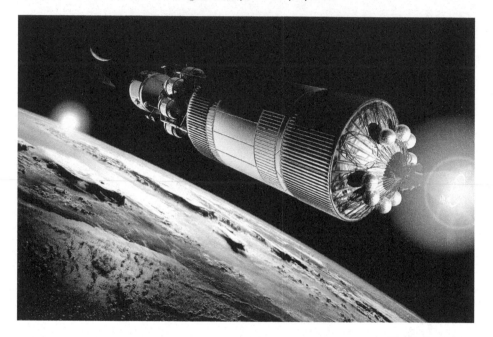

1.1 Q: On the surface of Earth, a spacecraft has a mass of 2.00×10^4 kg. What is the mass of the spacecraft at a distance of one Earth radius above Earth's surface?

(1) 5.00×10^3 kg

(2) 2.00×10^4 kg

(3) 4.90×10^4 kg

(4) 1.96×10^5 kg

1.1 A: (2) 2.00×10^4 kg. Mass is constant, therefore the spacecraft's mass at a distance of one Earth radius above Earth's surface is 2.00×10^4 kg.

Energy

If it's not matter, what's left? Why, energy, of course. As energy is such an everyday term that encompasses so many areas, an accurate definition can be quite elusive. Physics texts oftentimes define **energy** as the ability or capacity to do work. It's a nice, succinct definition, but leads to another question – what is work? **Work** can also be defined many ways, but let's start with the process of moving an object. If we put these two definitions together, we can vaguely define energy as the ability or capacity to move an object.

Mass – Energy Equivalence

So far, our definition of physics boils down to the study of matter, energy, and their interactions. Around the turn of the 20th century, however, several physicists began proposing a strong relationship between matter and energy. Albert Einstein, in 1905, formalized this with his famous formula $E=mc^2$, which relates that the mass of an object, a key characteristic of matter, is really a measure of its energy. This discovery has paved the way for tremendous innovation ranging from nuclear power plants to atomic weapons to particle colliders performing research on the origins of the universe. Ultimately, if traced back to its origin, the source of all energy on earth is the conversion of mass to energy!

Answering the Question

Physics, in some sense, can therefore be defined as the study of just about everything. Try to think of something that isn't physics – go on, I dare you! Not so easy, is it? Even the more ambiguous topics can, in some sense, be categorized as physics. A Shakespearean sonnet? A sonnet is typically read from a manuscript (matter), and sensed by the conversion of light (energy) alternately reflected and absorbed from a substrate, focused by a lens in the eye, and converted to chemical and electrical signals by photoreceptors on the retina, and then transferred as electrical and chemical signals along the neural pathways to the brain. That's just the high-level analysis. In short, just about everything is physics from a certain perspective.

As this is an introductory course in physics, we'll limit our scope somewhat as we build up a foundational understanding of the world around us. We'll begin with a study of Newtonian Mechanics, which explores moving objects and their interactions. From there, we'll move into electricity and magnetism. We'll then combine moving objects and electromagnetism as we dive into sound and physical waves. Then, using what we've learned about physical waves, we'll expand into electromagnetic waves and optics. Finally, we'll end the course by looking at matter again, this time at the nuclear level, using our background in mechanics, electricity and magnetism, waves, and optics to build a deeper understanding of the building blocks of our universe.

Join me as we take our first steps into a better understanding of the universe we live in.

Chapter 2: Math Review

"Whoever wants to understand much must play much."

— Gottfried Benn

Objectives

Review required concepts and skills required for success in the NY Regents Physics course.

1. Express answers correctly with respect to significant figures.
2. Use scientific notation to express physical values efficiently.
3. Convert and estimate SI units.
4. Differentiate between scalar and vector quantities.
5. Use scaled diagrams to represent and manipulate vectors.
6. Determine x- and y-components of two-dimensional vectors.
7. Determine the angle of a vector given its components.

Significant Figures

Significant Figures (or sig figs, for short) represent a manner of showing which digits in a number are known to some level of certainty. But how do we know which digits are significant? There are some rules to help us with this. If we start with a number in scientific notation:

- All non-zero digits are significant.
- All digits between non-zero digits are significant.
- Zeroes to the left of significant digits are not significant.
- Zeroes to the right of significant digits are significant.

When you make a measurement in physics, you want to write what you measured using significant figures. To do this, write down as many digits as you are absolutely certain of, then take a shot at one more digit as accurately as you can. These are your significant figures. On the Regents Physics Exam, you may answer any problem by showing three or four significant figures.

2.1 Q: How many significant figures are in the value 43.74 km?

2.1 A: 4 (four non-zero digits)

2.2 Q: How many significant figures are in the value 4302.5 g?

2.2 A: 5 (All non-zero digits are significant and digits between non-zero digits are significant.)

2.3 Q: How many significant figures are in the value 0.0083s?

2.3 A: 2 (All non-zero digits are significant. Zeroes to the left of significant digits are not significant.)

2.4 Q: How many significant figures are in the value 1.200×10^3 kg?

2.4 A: 4 (Zeroes to the right of significant digits are significant.)

Scientific Notation

Although physics and mathematics aren't the same thing, they are in many ways closely related. Just like English is the language of this content, mathematics is the language of physics. A solid understanding of a few simple math concepts will allow us to communicate and describe the physical world both efficiently and accurately.

Because measurements of the physical world vary so tremendously in size (imagine trying to describe the distance across the United States in units of hair thicknesses), physicists often times use what is known as scientific notation to denote very large and very small numbers. These very large and very small numbers would become quite cumbersome to write out repeatedly. Imagine writing 4,000,000,000,000 over and over again. Your hand would get tired and your pen would rapidly run out of ink! Instead, it's much easier to write this number as 4×10^{12}. See how much easier that is? Or on the smaller scale, the thickness of the insulating layer (known as a gate dielectric) in the integrated circuits that power our computers and other electronics can be less than 0.000000001 m. It's easy to lose track of how many zeros you have to deal with, so scientists instead would write this number as 1×10^{9} m. See how much simpler life can be with scientific notation?

Scientific notation follows these simple rules. Start by showing all the significant figures in the number you're describing, with the decimal point after the first significant digit. Then, show your number being multiplied by 10 to the appropriate power in order to give you the correct value.

It sounds more complicated than it is. Let's say, for instance, you want to show the number 300,000,000 in scientific notation (a very useful number in physics), and let's assume we know this value to three significant digits. We would start by writing our three significant digits, with the decimal point after the first digit, as "3.00". Now, we need to multiply this number by 10 to some power in order to get back to our original value. In this case, we multiply 3.00 by 10^{8}, for an answer of 3.00×10^{8}. Interestingly, the power you raise the 10 to is exactly equal to the number of digits you moved the decimal to the left as you converted from standard to scientific notation. Similarly, if you start in scientific notation, to convert to standard notation, all you have to do is remove the 10^{8} power by moving the decimal point eight digits to the right. Presto, you're an expert in scientific notation!

But, what do you do if the number is much smaller than one? Same basic idea... let's assume we're dealing with the approximate radius of an electron, which is 0.00000000000000282 m. It's easy to see how unwieldy this could become. We can write this in scientific notation by writing our three significant digits, with the decimal point after the first digit, as "2.82." Again, we multiply this number by some power to 10 in order to get back to our original value. Because our value is less than 1, we need to use negative powers of 10. If we raise 10 to the power -15, specifically, we get a final value of 2.82×10^{-15} m. In essence, for every digit we moved the decimal place, we add another power of 10. And if we start with scientific notation, all we do is move the decimal place left one digit for every negative power of 10.

2.5 Q: Express the number 0.000470 in scientific notation.
2.5 A: 4.70×10^{-4}

2.6 Q: Express the number 2,870,000 in scientific notation.
2.6 A: 2.87×10^{6}

2.7 Q: Expand the number 9.56×10^{-3}.
2.7 A: 0.00956

2.8 Q: Expand the number 1.11×10^{7}.
2.8 A: 11,100,000

Metric System

Physics involves the study, prediction, and analysis of real-world phenomena. To communicate data accurately, we must set specific standards for our basic measurements. The physics community has standarized on what is known as the **Système International** (SI), which defines seven baseline measurements and their standard units, forming the foundation of what is called the metric system of measurement. The SI system is oftentimes referred to as the mks system, as the three most common measurement units are meters, kilograms, and seconds, which we'll focus on for the majority of this course. The fourth SI base unit we'll use in this course, the ampere, will be introduced in the current electricity section.

The base unit of length in the metric system, the meter, is roughly equivalent to the English yard. For smaller measurements, the meter is divided up into 100 parts, known as centimeters, and each centimeter is made up of 10 millimeters. For larger measurements, the meter is grouped into larger units of 1000 meters, known as a kilometer. The length of a baseball bat is approximately one meter, the radius of a U.S. quarter is approximately a centimeter, and the diameter of the metal in a wire paperclip is roughly one millimeter.

The base unit of mass, the kilogram, is roughly equivalent to two U.S. pounds. A cube of water 10cm x 10cm x 10cm has a mass of 1 kilogram. Kilograms can also be broken up into larger and smaller units, with commonly used measurements of grams (1/1000th of a kilogram) and milligrams (1/1000th of a gram). The mass of a textbook is approximately 2 to 3 kilograms, the mass of a baseball is approximately 145 grams, and the mass of a mosquito is 1 to 2 grams.

The base unit of time, the second, is likely already familiar. Time can also be broken up into smaller units such as milliseconds (10^{-3} seconds), microseconds (10^{-6} seconds), and nanoseconds (10^{-9} seconds), or grouped into larger units such as minutes (60 seconds), hours (60 minutes), days (24 hours), and years (365.25 days).

The metric system is based on powers of 10, allowing for easy conversion from one unit to another. The front page of the Physics Reference Table includes a chart showing the meaning of the commonly used metric prefixes, which can be extremely valuable in performing unit conversions.

Converting from one unit to another can be accomplished in a straightforward manner if you follow the described procedure:

1. Write your initial measurement with units as a fraction over 1.
2. Multiply your initial fraction by a second fraction, with a numerator (top number) having the units you want to convert to, and the denominator (bottom number) having the units of your

initial measurement.
3. For any units on the top right-hand side with a prefix, use the reference table to determine the value for that prefix. Write that prefix in the right-hand denominator. If there is no prefix, use 1.
4. For any units on the right-hand denominator with a prefix, write the value for that prefix in the right-hand numerator. If there is no prefix, use 1.
5. Multiply through the problem, taking care to accurately record units. You should be left with a final answer in the desired units.

Let's take a look at a sample unit conversion:

2.9 Q: Convert 23 millimeters (mm) to meters (m).

2.9 A: Step 1. $\dfrac{23mm}{1}$

Step 2. $\dfrac{23mm}{1} \times \dfrac{m}{mm}$

Step 3. $\dfrac{23mm}{1} \times \dfrac{m}{1mm}$

Step 4. $\dfrac{23mm}{1} \times \dfrac{10^{-3}m}{1mm}$

Step 5. $\dfrac{23mm}{1} \times \dfrac{10^{-3}m}{1mm} = 2.3 \times 10^{-2} m$

Now let's have you try some on your own...

2.10 Q: Convert 2.67×10^{-4} m to mm.

2.10 A: $\dfrac{2.67 \times 10^{-4} m}{1} \times \dfrac{1mm}{10^{-3}m} = 0.267 mm$

2.11 Q: Convert 14kg to mg.

2.11 A: $\dfrac{14kg}{1} \times \dfrac{10^{3}mg}{10^{-3}kg} = 14 \times 10^{6} mg$

2.12 Q: Convert 3,470,000 µs to s.

2.12 A: $\dfrac{3,470,000\mu s}{1} \times \dfrac{10^{-6}s}{1\mu s} = 3.47s$

2.13 Q: Convert 64GB to KB and express in scientific notation.

2.13 A: $\dfrac{64GB}{1} \times \dfrac{10^{9}\,KB}{10^{3}\,GB} = 64,000,000\,KB = 6.4 \times 10^{7}\,KB$

Algebra and Trigonometry

Just as we find the English language a convenient tool for conveying thoughts to each other,
we need a convenient language
for conveying our understanding
of the world around us in order
to understand its behavior. The
language most commonly (and
conveniently) used to describe
the natural world is mathematics.
Therefore, to understand physics, we need to be fluent in the
mathematics of the topics we'll
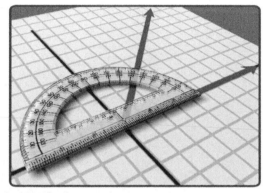
study in this course... specifically basic algebra and trigonometry.

Now don't you fret or frown, for those whom the word "trig" conjures up
feelings of pain, angst, and frustration, have no fear. We will need only the
most basic of algebra and trigonometry knowledge in order to successfully
solve a wide range of physics problems.

A vast majority of problems requiring algebra can be solved using the same
problem solving strategy. First, analyze the problem and write down what
you know, what you need to find, and make a picture or diagram to better
understand the problem if it makes sense. Then, start your solution by se-
lecting an appropriate formula from the Reference Table (a comprehensive
formula sheet covering material for the entire course). Next, substitute given
information from the problem statement into the formula (with units). Finally,
solve the problem for your final answer, making sure to show your answer
with units. This strategy, known as the AFSA strategy, can be summarized as:

- Analysis – Write down what is given, what you're asked to find,
  and make a diagram, if appropriate.
- Formula – Copy the appropriate formula from the reference table.
- Substitution (with units) – Replace variables in the selected
  formula with data from the problem statement.
- Answer (with units) – Solve the problem and record your final
  answer with units. Place a box around your final answer.

Our use of trigonometry, the study of triangles, can be distilled down into the definitions of the three basic trigonometric functions. If you can use the definitions of the sine, cosine, and tangent, you'll be fine in this course. Even better – if you're a little rusty on the basic trig definitions, they're all included for you on the Regents Physics Reference Table – you don't even have to memorize them.

Remember to use the AFSA strategy and your reference table and you'll be well prepared for anything the Regents Physics Exam may throw at you.

Vectors and Scalars

Quantities in physics are used to represent real-world measurements, and therefore physicists use these quantities as tools to better understand the world. In examining these quantities, there are times when just a number, with a unit, can completely describe a situation. These numbers, which have just a **magnitude**, or size, are known as **scalars**. Examples of scalars include quantities such as temperature, mass, and time. At other times, a quantity is more descriptive if it also includes a direction. These quantities which have both a magnitude and direction are known as **vectors**. Vector quantities you may be familiar with include force, velocity, and acceleration.

Most students will be familiar with scalars, but to many, vectors may be a new and confusing concept. By learning just a few rules for dealing with vectors, you'll find that they can be a powerful tool for problem solving.

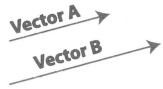

Vectors are often represented as arrows, with the length of the arrow indicating the magnitude of the quantity, and the direction of the arrow indicating the direction of the vector. In the figure at left, vector B has a magnitude greater than that of vector A even though vectors A and B point in the same direction. It's also important to know that vectors can be moved anywhere in space. The positions of A and B could be reversed, and the individual vectors would retain their values of magnitude and direction.

To add vectors A and B all we have to do is line them up so that the tip of the first vector touches the tail of the second vector. Then, to find the sum of the vectors, known as the resultant, all we have to do is draw a straight line from the start of the first vector to the end of the last vector. This method works with any number of vectors.

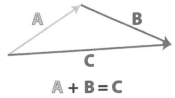

$A + B = C$

So how do we subtract two vectors? Let's try it by subtracting B from A. We could rewrite the expression A - B as A + -B. Now it becomes an addition problem, we just have to figure out how to express –B. This is easier than it sounds. To find the opposite of a vector, we just point the vector in the opposite direction. Therefore, we can use what we already know about the addition of vectors to find the resultant of A-B.

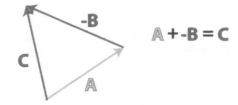

A + -B = C

Components of Vectors

We'll learn more about vectors as we go, but before we move on, there are a few basic skills we need to learn. Vectors at angles can be challenging to deal with. By transforming a vector at an angle into two vectors, one parallel to the x-axis and one parallel to the y-axis, we can greatly simplify problem solving. To break a vector up into its components, we can use our basic trig functions. To help us out even further, the Regents Physics Reference Table includes the exact formulas we need to determine the x- and y-components of any vector if we know that vector's magnitude and direction.

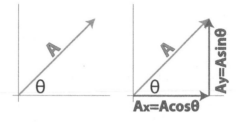

2.14 Q: The vector diagram below represents the horizontal component, F_H, and the vertical component, F_V, of a 24-Newton force acting at 35° above the horizontal. What are the magnitudes of the horizontal and vertical components?

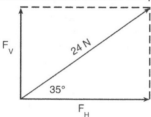

(1) F_H=3.5 N and F_V=4.9 N
(2) F_H=4.9 N and F_V=3.5 N
(3) F_H=14 N and F_V=20 N
(4) F_H=20 N and F_V=14 N

2.14 A: (4) $F_H = A_x = A\cos\theta = (24N)\cos 35° = 20N$

$F_V = A_y = A\sin\theta = (24N)\sin 35° = 14N$

2.15 Q: An airplane flies with a velocity of 750 kilometers per hour, 30° south of east. What is the magnitude of the plane's eastward velocity?

(1) 866 km/h

(2) 650 km/h

(3) 433 km/h

(4) 375 km/h

2.15 A:

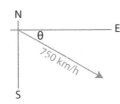

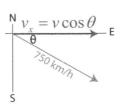

(2) $v_x = v\cos\theta = (750\,{}^{km}\!/_h)\cos(30°) = 650\,{}^{km}\!/_h$

2.16 Q: A soccer player kicks a ball with an initial velocity of 10 m/s at an angle of 30° above the horizontal. The magnitude of the horizontal component of the ball's velocity is

(1) 5.0 m/s

(2) 8.7 m/s

(3) 9.8 m/s

(4) 10 m/s

2.16 A:

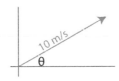

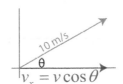

(2) $v_x = v\cos\theta = (10\,{}^{m}\!/_s)\cos(30°) = 8.7\,{}^{m}\!/_s$

2.17 Q: A child kicks a ball with an initial velocity of 8.5 meters per second at an angle of 35° with the horizontal, as shown. The ball has an initial vertical velocity of 4.9 meters per second. The horizontal component of the ball's initial velocity is approximately

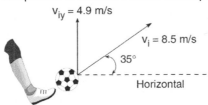

(1) 3.6 m/s

(2) 4.9 m/s

(3) 7.0 m/s

(4) 13 m/s

2.17 A: (3) $v_x = v\cos\theta = (8.5\,{}^m/_s)\cos(35°) = 6.96\,{}^m/_s$

In similar fashion, we can use the components of a vector in order to build the original vector. Graphically, if we line up the component vectors tip-to-tail, the original vector runs from the starting point of the first vector to the ending point of the last vector. To determine the magnitude of the resulting vector algebraically, just apply the Pythagorean Theorem.

2.18 Q: A motorboat, which has a speed of 5.0 meters per second in still water, is headed east as it crosses a river flowing south at 3.3 meters per second. What is the magnitude of the boat's resultant velocity with respect to the starting point?

(1) 3.3 m/s

(2) 5.0 m/s

(3) 6.0 m/s

(4) 8.3 m/s

2.18 A: (3) 6.0 m/s

The motorboat's resultant velocity is the vector sum of the motorboat's speed and the riverboat's speed.

$$a^2 + b^2 = c^2$$
$$c = \sqrt{a^2 + b^2}$$
$$c = \sqrt{(5\,{}^m/_s)^2 + (3.3\,{}^m/_s)^2}$$
$$c = 6\,{}^m/_s$$

2.19 Q: A dog walks 8.0 meters due north and then 6.0 meters due east. Determine the magnitude of the dog's total displacement.

2.19 A:
$$a^2 + b^2 = c^2$$
$$c = \sqrt{a^2 + b^2}$$
$$c = \sqrt{(6\,{}^m/_s)^2 + (8\,{}^m/_s)^2}$$
$$c = 10\,{}^m/_s$$

Chapter 2: Math Review

2.20 Q: A 5.0-newton force could have perpendicular components of
(1) 1.0 N and 4.0 N
(2) 2.0 N and 3.0 N
(3) 3.0 N and 4.0 N
(4) 5.0 N and 5.0 N

2.20 A: (3) The only answers that fit the Pythagorean Theorem are 3.0 N and 4.0 N ($3^2+4^2=5^2$)

2.21 Q: A vector makes an angle, θ, with the horizontal. The horizontal and vertical components of the vector will be equal in magnitude if angle θ is
(1) 30°
(2) 45°
(3) 60°
(4) 90°

2.21 A: (2) 45°. $A_x=A\cos(\theta)$ will be equal to $A_y=A\sin(\theta)$ when angle θ=45° since $\cos(45°)=\sin(45°)$.

The Equilibrant Vector

The equilibrant of a force vector or set of force vectors is a single force vector which is exactly equal in magnitude and opposite in direction to the original vector or sum of vectors. The equilibrant, in effect, "cancels out" the original vector(s), or brings the set of vectors into equilibrium. To find an equilibrant, first find the resultant of the original vectors. The equilibrant is the opposite of the resultant you found!

2.22 Q: The diagram below represents two concurrent forces.

Which vector represents the force that will produce equilibrium with these two forces?

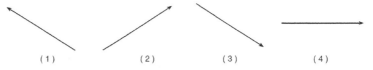

| (1) | (2) | (3) | (4) |

2.22 A: (3) The resultant of the two vectors would point up and to the left, therefore the equilibrant must point in the opposite direction, down and to the right.

Chapter 3: Kinematics

"I like physics. I think it is the best science out of all three of them, because generally it's more useful. You learn about speed and velocity and time, and that's all clever stuff."

— Tom Felton

Objectives

1. Understand the difference between distance and displacement and between speed and velocity.
2. Calculate distance, displacement, speed, velocity, and acceleration.
3. Solve problems involving average speed and average velocity.
4. Construct and interpret graphs of position, velocity, and acceleration versus time.
5. Determine and interpret slopes and areas of motion graphs.
6. Determine the acceleration due to gravity near the surface of Earth.
7. Use kinematic equations to solve problems for objects moving at a constant acceleration in a straight line and in free fall.
8. Resolve a vector into perpendicular components: both graphically and algebraically.
9. Sketch the theoretical path of a projectile.
10. Recognize the independence of the vertical and horizontal motions of a projectile.
11. Solve problems involving projectile motion for projectiles fired horizontally and at an angle.

Regents Physics is all about energy in the universe, in all its various forms. Here on Earth, the source of our energy, directly or indirectly, is the sun. Solar power, wind power, hydroelectric power, fossil fuels, we can eventually trace the origin of all energy on our planet back to our sun. So where do we start in our study of the universe?

Theoretically, we could start by investigating any of these types of energy. In reality, however, by starting with energy of motion (also known as **kinetic energy**), we can develop a set of analytical problem solving skills from basic principles that will serve us well as we expand into our study of other types of energy.

For an object to have kinetic energy, it must be moving. Specifically, the kinetic energy of an object is equal to one half times the object's mass multiplied by the square of its velocity.

$$KE = \tfrac{1}{2} mv^2$$

If kinetic energy is energy of motion, and energy is the ability or capacity to do work (moving an object), then we can think of kinetic energy as the ability or capacity of a moving object to move another object.

But what does it mean to be in motion? A moving object has a varying position... its location changes as a function of time. So to understand kinetic energy, we'll need to better understand position and how position changes. This will lead us into our first major unit, kinematics, from the Greek word kinein, meaning to move. Formally, kinematics is the branch of physics dealing with the description of an object's motion, leaving the study of the "why" of motion to our next major topic, dynamics.

Chapter 3: Kinematics

Distance and Displacement

An object's **position** refers to its location at any given point in time. If we confine our study to motion in one dimension, we can define how far an object travels from its initial position as its **distance**, "d." Distance, as defined by physics, is a scalar. It has a magnitude, or size, only. The basic unit of distance is the meter (m).

3.1 Q: On a sunny afternoon, a deer walks 1300 meters east to a creek for a drink. The deer then walks 500 meters west to the berry patch for dinner, before running 300 meters west when startled by a loud raccoon. What distance did the deer travel?

3.1 A: The deer traveled 1300m + 500m + 300m, for a total distance traveled of 2100m.

Besides distance, in physics it is also helpful to know how far an object is from its starting point. The vector quantity **displacement** describes how far an object is from its starting point, and the direction of the displacement vector points from the starting point to the finishing point. Like distance, the units of displacement are meters. Complicating matters, though, displacement also uses the same symbol as distance, a "d."

3.2 Q: A deer walks 1300 m east to a creek for a drink. The deer then walked 500 m west to the berry patch for dinner, before running 300 m west when startled by a loud raccoon. What is the deer's displacement?

3.2 A: The deer's displacement was 500m east.

3.3 Q: Which is a vector quantity?
(1) speed
(2) work
(3) mass
(4) displacement

3.3 A: (4) Displacement is a vector quantity; it has direction.

3.4 Q: A student on her way to school walks four blocks east, three blocks north, and another four blocks east, as shown in the diagram.

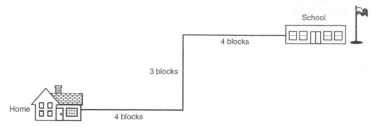

Compared to the distance she walks, the magnitude of her displacement from home to school is

(1) less

(2) greater

(3) the same

3.4 A: (1) The magnitude of displacement is always less than or equal to the distance traveled.

3.5 Q: A hiker walks 5 kilometers due north and then 7 kilometers due east. What is the magnitude of her resultant displacement? What total distance has she traveled?

3.5 A:

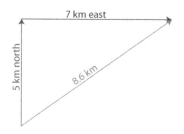

$$a^2 + b^2 = c^2 \rightarrow c = \sqrt{a^2 + b^2} = \sqrt{(5km)^2 + (7km)^2} = 8.6km$$

The hiker's resultant displacement is 8.6 km north of east

The hiker's distance traveled is 12 kilometers.

Notice how for the exact same motion, distance and displacement have significantly different values based on their scalar or vector nature. Understanding the similarities (and differences) between these concepts is an important step toward understanding kinematics.

Speed and Velocity

Knowing only an object's distance and displacement doesn't tell us the whole story. Going back to our deer example, there's a significant difference in our picture of the deer's afternoon if the deer's travels occurred over 5 minutes (300 seconds) as opposed to over 50 minutes (3000 seconds).

How exactly does the picture change? In order to answer that question, we'll need to introduce some new concepts – average speed and average velocity. Both physics quantities use the same formula and the same units (m/s), but the symbols stand for different things in each formula. In this way, you can re-use the same formula to give you two different physics quantities. The formula, available on the back page of your reference table, states that average velocity is displacement divided by time. The formula also states that average speed equals distance divided by time:

$$\overline{v} = \frac{d}{t}$$

Average speed, given the symbol $\overline{v}$, is defined as distance divided by time, and it tells you how quickly an object's distance changes. To calculate the scalar quantity average speed, you divide the scalar quantity distance by time.

3.6 Q: A deer walks 1300 m east to a creek for a drink. The deer then walked 500 m west to the berry patch for dinner, before running 300 m west when startled by a loud raccoon. What is the deer's average speed if the entire trip took 600 seconds (10 minutes)?

3.6 A: $\overline{v} = \dfrac{d}{t} = \dfrac{2100m}{600s} = 3.5 \,^{m}\!/_{s}$

Average velocity, also given the symbol $\overline{v}$, is defined as displacement over time. It tells you how quickly an object's displacement changes. To calculate the vector quantity average velocity, you divide the vector quantity displacement by time.

3.7 Q: A deer walks 1300 m east to a creek for a drink. The deer then walked 500 m west to the berry patch for dinner, before running 300 m west when startled by a loud raccoon. What is the deer's average velocity if the entire trip took 600 seconds (10 minutes)?

3.7 A: $\overline{v} = \dfrac{d}{t} = \dfrac{500m}{600s} = 0.83 \,^{m}\!/_{s}$ east

Again, notice how you get very different answers for average speed compared to average velocity. The difference is realizing that distance and speed are scalars, and displacement and velocity are vectors. One way to help you remember these: **s**peed is a **s**calar, and **v**elocity is a **v**ector.

3.8 Q: Chuck the hungry squirrel travels 4m east and 3m north in search of an acorn. The entire trip takes him 20 seconds. Find: Chuck's distance traveled, Chuck's displacement, Chuck's average speed, and Chuck's average velocity.

3.8 A:

$$d \text{ (distance)} = 4m + 3m = 7m$$

$$d \text{ (displacement)} = \sqrt{(4m)^2 + (3m)^2}$$

$$d \text{ (displacement)} = 5m \text{ } northeast$$

$$\overline{v} \text{ (avg. speed)} = \frac{d}{t} = \frac{7m}{20s} = 0.35 \text{ }^m/_s$$

$$\overline{v} \text{ (avg. velocity)} = \frac{d}{t} = \frac{5m}{20s} = 0.25 \text{ }^m/_s \text{ } northeast$$

Let's try a few more problems to demonstrate the potential applications of these definitions.

3.9 Q: On a highway, a car is driven 80 kilometers during the first 1.00 hour of travel, 50 kilometers during the next 0.50 hour, and 40 kilometers in the final 0.50 hour. What is the car's average speed for the entire trip?

(1) 45 km/h
(2) 60 km/h
(3) 85 km/h
(4) 170 km/h

3.9 A: (3) $\overline{v} = \frac{d}{t} = \frac{170km}{2h} = 85 \text{ }^{km}/_h$

3.10 Q: A person walks 150 meters due east and then walks 30 meters due west. The entire trip takes the person 10 minutes. Determine the magnitude and the direction of the person's total displacement.

3.10 A: 120m due east

Acceleration

So we're starting to get a pretty good understanding of motion. But what would our world be like if velocity never changed? Objects at rest would remain at rest. Objects in motion would remain in motion at a constant speed and direction. And kinetic energy would never change (remember $KE = \frac{1}{2}mv^2$?) It'd make for a pretty boring world. Thankfully, velocity can change in our world, and we call a change in velocity an **acceleration**.

More accurately, acceleration is the rate at which velocity changes. We can write this as:

$$a = \frac{\Delta v}{t}$$

This indicates that the change in velocity divided by the time interval gives you the acceleration (another of our kinematic equations on the reference table). Much like displacement and velocity, acceleration is a vector – it has a direction. Further, the units of acceleration are meters per second per second, or [m/s²]. Although it sounds complicated, all the units mean is that velocity changes at the rate of one meter per second, every second. So an object starting at rest and accelerating at 2 m/s² would be moving at 2 m/s after one second, 4 m/s after two seconds, 6 m/s after three seconds, and so on.

Of special note is the symbolism for Δv. The delta symbol (Δ) indicates a change in a quantity, which is always the initial quantity subtracted from the final quantity. For example:

$$\Delta v = v_f - v_i$$

3.11 Q: Monty the Monkey accelerates uniformly from rest to a velocity of 9 m/s in a time span of 3 seconds. Calculate Monty's acceleration.

3.11 A: $a = \dfrac{\Delta v}{t} = \dfrac{v_f - v_i}{t} = \dfrac{9\,m/s - 0\,m/s}{3s} = 3\,m/s^2$

3.12 Q: Velocity is to speed as displacement is to
(1) acceleration
(2) time
(3) momentum
(4) distance

3.12 A: (4) distance. Velocity is the vector equivalent of speed, and displacement is the vector equivalent of distance.

To develop another of our kinematic equations, we can combine the definition of acceleration and the change in velocity equation as follows:

$$a = \frac{\Delta v}{t} = \frac{v_f - v_i}{t}$$

$$v_f - v_i = at$$

$$v_f = v_i + at$$

3.13 Q: The instant before a batter hits a 0.14-kilogram baseball, the velocity of the ball is 45 meters per second west. The instant after the batter hits the ball, the ball's velocity is 35 meters per second east. The bat and ball are in contact for 1.0×10^{-2} second. Determine the magnitude and direction of the average acceleration of the baseball while it is in contact with the bat.

3.13 A:

Given: Find:

$v_i = -45 \,{}^m\!/_s$ a

$v_f = 35 \,{}^m\!/_s$

$t = 1 \times 10^{-2} s$

$$a = \frac{\Delta v}{t} = \frac{v_f - v_i}{t} = \frac{35\,{}^m\!/_s - (-45\,{}^m\!/_s)}{1 \times 10^{-2}\, s}$$

$a = 8000 \,{}^m\!/_{s^2}$ east

Because acceleration is a vector and has direction, it's important to realize that positive and negative values for acceleration indicate direction only. Let's take a look at some examples. First, an acceleration of zero implies an object moves at a constant velocity, so a car traveling at 30 m/s east with zero acceleration remains in motion at 30 m/s east.

If the car starts at rest and the car is given a positive acceleration of 5 m/s² east, the car speeds up as it moves east, going faster and faster each second. After one second, the car is traveling 5 m/s. After two seconds, the car travels 10 m/s. After three seconds, the car travels 15 m/s, and so on.

But what happens if the car starts with a velocity of 15 m/s east, and it accelerates at a rate of -5 m/s² (or equivalently, 5 m/s² west)? The car will slow down as it moves to the east until its velocity becomes zero, then it will speed up as it continues to the west.

Positive accelerations don't necessarily indicate an object speeding up, and negative accelerations don't necessarily indicate an object slowing down. In one dimension, for example, if we call east the positive direction, a negative acceleration would indicate an acceleration vector pointing to the west. If the object is moving to the east (has a positive velocity), the negative acceleration would indicate the object is slowing down. If, however, the object is moving to the west (has a negative velocity), the negative acceleration would indicate the object is speeding up as it moves west.

Exasperating, isn't it? Putting it much more simply, if acceleration and velocity have the same sign (vectors in the same direction), the object is speeding up. If acceleration and velocity have opposite signs (vectors in opposite directions), the object is slowing down.

Particle Diagrams

Graphs and diagrams are terrific tools for understanding physics, and they are especially helpful for studying motion, a phenomenon that we are used to perceiving visually. Particle diagrams, sometimes referred to as ticker-tape diagrams or dot diagrams, show the position or displacement of an object at evenly spaced time intervals.

Think of a particle diagram like an oil drip pattern... if a car has a steady oil drip, where one drop of oil falls to the ground every second, the pattern of the oil droplets on the ground could represent the motion of the car with respect to time. By examining the oil drop pattern, a bystander could draw conclusions about the displacement, velocity, and acceleration of the car, even if they weren't able to watch the car drive by! The oil drop pattern is known as a particle, or ticker-tape, diagram.

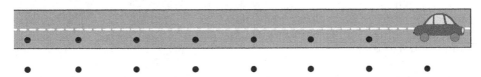

From the particle diagram above we can see that the car was moving either to the right or the left, and since the drops are evenly spaced, we can say with certainty that the car was moving at a constant velocity, and since velocity isn't changing, acceleration must be 0. So what would the particle diagram look like if the car was accelerating to the right? Let's take a look and see!

The oil drops start close together on the left, and get further and further apart as the object moves toward the right. Of course, this pattern could also have been produced by a car moving from right to left, beginning with a high velocity at the right and slowing down as it moves toward the left. Because the velocity vector (pointing to the left) and the acceleration vector (pointing to the right) are in opposite directions, the object slows down. This is a case where, if you called to the right the positive direction, the car would have a negative velocity, a positive acceleration, and it would be slowing down. Check out the resulting particle diagram below!

● ● ● ● ● ● ●

Can you think of a case in which the car could have a negative velocity and a negative acceleration, yet be speeding up? Draw a picture of the situation!

Displacement-Time (d-t) Graphs

As you've observed, particle diagrams can help you understand an object's motion, but they don't always tell you the whole story. We'll have to investigate some other types of motion graphs to get a clearer picture.

The displacement time graph (also known as a d-t graph or position-time graph) shows the displacement (or, in the case of scalar quantities, distance) of an object as a function of time. Positive displacements indicate the object's position is in the positive direction from its starting point, while negative displacements indicate the object's position is opposite the positive direction. Let's look at a few examples.

Suppose Cricket the Wonder Dog wanders away from her house at a constant velocity of 1 m/s, stopping only when she's 5m away (which, of course, takes 5 seconds). She then decides to take a short five second rest in the grass. After her five second rest, she hears the dinner bell ring, so she runs back to the house at a speed of 2 m/s. The displacement-time graph for her motion would look something like this:

As you can see from the plot, Cricket's displacement begins at zero meters at time zero. Then, as time progresses, Cricket's displacement increases at a rate of 1 m/s, so that after one second, Cricket is one meter away from

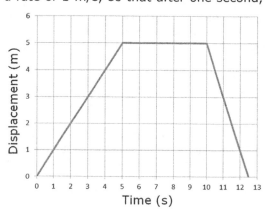

her starting point. After two seconds, she's two meters away, and so forth, until she reaches her maximum displacement of five meters from her starting point at a time of five seconds. Cricket then remains at that position for 5 seconds while she takes a rest. Following her rest, at time t=10 seconds, Cricket hears the dinner bell and races back to the house at a speed of 2 m/s, so the graph ends when Cricket returns to her starting point at the house, a total distance traveled of 10m, and a total displacement of zero meters.

As we look at the d-t graph, notice that at the beginning, when Cricket is moving in a positive direction, the graph has a positive slope. When the graph is flat (has a zero slope), Cricket is not moving. When the graph has a negative slope, Cricket is moving in the negative direction. It's also easy to see that the steeper the slope of the graph, the faster Cricket is moving.

3.14 Q: The graph below represents the displacement of an object moving in a straight line as a function of time.

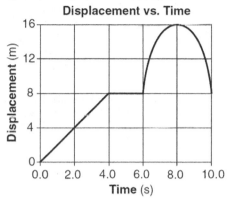

What was the total distance traveled by the object during the 10-second time interval?

(1) 0 m

(2) 8 m

(3) 16 m

(4) 24 m

3.14 A: (4) Total distance traveled is 8 meters forward from 0 to 4 seconds, then 8 meters forward from 6 to 8 seconds, then 8 meters backward from 8 to 10 seconds, for a total of 24 meters.

Velocity-Time (v-t) Graphs

Just as important to understanding motion is the velocity-time graph, which shows the velocity of an object on the y-axis, and time on the x-axis. Positive values indicate velocities in the positive direction, while negative values indicate velocities in the opposite direction. In reading these graphs, it's important to realize that a straight horizontal line indicates the object maintaining a constant velocity – it can still be moving, it's velocity just isn't changing. A value of 0 on the v-t graph indicates the object has come to a stop. If the graph crosses the x-axis, the object was moving in one direction, came to a stop, and switched the direction of its motion. Let's look at the v-t graph for Cricket the Wonderdog's Adventure from the d-t graph section:

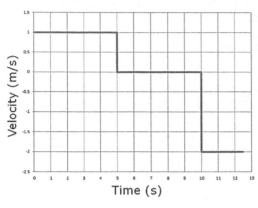

Velocity (m/s) vs Time (s)

For the first five seconds of Cricket's journey, we can see she maintains a constant velocity of 1 m/s. Then, when she stops to rest, her velocity changes to zero for the duration of her rest. Finally, when she races back to the house for dinner, she maintains a negative velocity of 2 m/s. Because velocity is a vector, the negative sign indicates that Cricket's velocity is in the opposite direction (we initially defined the direction away from the house as positive, so back toward the house must be negative!)

As I'm sure you can imagine, the d-t graph of an object's motion and the v-t graph of an object's motion are closely related. We'll explore these relationships next.

Graph Transformations

In looking at a d-t graph, the faster an object's displacement changes, the steeper the slope of the line. Since velocity is the rate at which an object's displacement changes, the slope of the d-t graph at any given point in time gives you the velocity at that point in time. We can obtain the slope of the d-t graph using the following formula:

$$slope = \frac{rise}{run} = \frac{y_2 - y_1}{x_2 - x_1}$$

Realizing that the rise in our graph is actually Δd, and the run is Δt, we can substitute these variables into our slope equation to find:

$$slope = \frac{rise}{run} = \frac{\Delta d}{\Delta t} = v$$

With a little bit of interpretation, it's easy to show that our slope is really just displacement over time, which is the definition of velocity. Put directly, the slope of the d-t graph is the velocity.

Of course, it only makes sense that if you can determine velocity from the d-t graph, you should be able to work backward to determine displacement from the v-t graph. If you have a v-t graph, and you want to know how much an object's displacement has changed in a time interval, take the area under the curve within that time interval.

So, if taking the slope of the d-t graph gives you the rate of change of displacement, which we call velocity, what do you get when you take the slope of the v-t graph? You get the rate of change of velocity, which we call acceleration! The slope of the v-t graph, therefore, tells you the acceleration of an object.

$$slope = \frac{rise}{run} = \frac{\Delta v}{\Delta t} = a$$

3.15 Q: The graph below represents the motion of a car during a 6.0-second time interval.

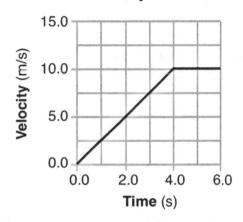

(A) What is the total distance traveled by the car during this 6-second interval?

(B) What is the acceleration of the car at t = 5 seconds?

3.15 A: (A) distance = area under graph

distance = $Area_{triangle}$ + $Area_{rectangle}$

distance = $\frac{1}{2}bh + lw$

distance = $\frac{1}{2}(4s)(10\,{}^{m}\!/_{s}) + (2s)(10\,{}^{m}\!/_{s})$

distance = $40m$

(B) acceleration = slope at t=5 seconds = 0 because graph is flat at t=5 seconds.

3.16 Q: The graph below represents the velocity of an object traveling in a straight line as a function of time.

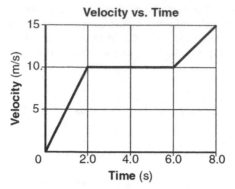

Velocity vs. Time

Determine the magnitude of the total displacement of the object at the end of the first 6.0 seconds.

3.16 A: displacement = area under graph

displacement = $Area_{triangle} + Area_{rectangle}$

displacement = $\frac{1}{2}bh + lw$

displacement = $\frac{1}{2}(2s)(10\,^{m}/_{s}) + (4s)(10\,^{m}/_{s})$

displacement = $50m$

3.17 Q: The graph below shows the velocity of a race car moving along a straight line as a function of time.

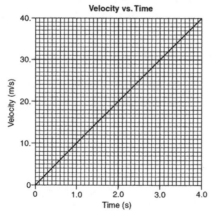

Velocity vs. Time

What is the magnitude of the displacement of the car from t = 2.0 seconds to t = 4.0 seconds?

(1) 20 m
(2) 40 m
(3) 60 m
(4) 80 m

3.17 A: (3) 60 m

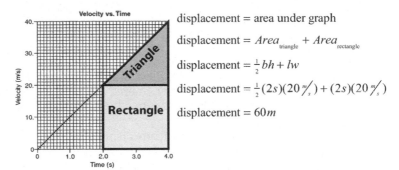

displacement = area under graph

displacement = $Area_{triangle} + Area_{rectangle}$

displacement = $\frac{1}{2}bh + lw$

displacement = $\frac{1}{2}(2s)(20\frac{m}{s}) + (2s)(20\frac{m}{s})$

displacement = $60m$

3.18 Q: The displacement-time graph below represents the motion of a cart initially moving forward along a straight line.

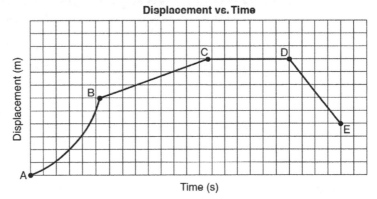

During which interval is the cart moving forward at constant speed?

(1) AB
(2) BC
(3) CD
(4) DE

3.18 A: (2) The slope of the d-t graph is constant and positive during interval BC, therefore the velocity of the cart must be constant and positive in that interval.

Acceleration-Time (a-t) Graphs

Much like we did with velocity, we can make a plot of acceleration vs. time by plotting the rate of change of an object's velocity (its acceleration) on the y-axis, and placing time on the x-axis. For the purposes of the NY Regents Physics Course, we'll always deal with a constant acceleration – all of our graphs will be a straight horizontal line, either at a positive value, a negative

---

value, or at 0 (indicating a constant velocity). It's important to understand, however, that in real life not all accelerations are constant.

When we took the slope of the d-t graph, we obtained an object's velocity. In the same way, taking the slope of the v-t graph gives you an object's acceleration. Going the other direction, when we analyzed the v-t graphs, we found that taking the area under the v-t graph provided us with information about the object's change in displacement. In similar fashion, taking the area under the a-t graph tells us how much an object's velocity changes.

Putting it all together, we can go from displacement-time to velocity-time by taking the slope, and we can go from velocity-time to acceleration-time by taking the slope. Or, going the other direction, the area under the acceleration-time curve gives you an object's change in velocity, and the area under the velocity-time curve gives you an object's change in displacement.

Graphical Analysis of Motion
How do I move from one type of graph to another?

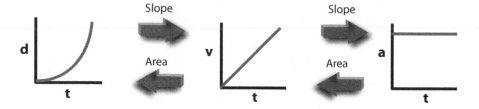

3.19 Q: Which graph best represents the motion of a block accelerating uniformly down an inclined plane?

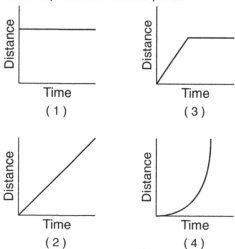

3.19 A: (4) If the block accelerates uniformly, it must have a constant acceleration. This means the v-t graph must be a straight line, since its slope, which is equal to its acceleration, must be constant. Therefore, the v-t graph must look something like the graph at right. The slope of the d-t graph, which gives velocity, must be constantly increasing to give the v-t graph above. The only answer choice with a constantly increasing slope is (4), so (4) must be the answer!

3.20 Q: A student throws a baseball vertically upward and then catches it. If vertically upward is considered to be the positive direction, which graph best represents the relationship between velocity and time for the baseball?

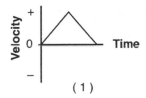

(1)

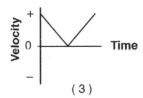

(3)

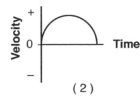

(2)

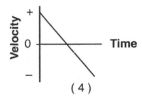

(4)

3.20 A: (4) If up is considered the positive direction, and the baseball is thrown upward to start its motion, it begins with a positive velocity. At its highest point, its vertical velocity becomes 0. Then, it speeds up as it comes down, so it obtains a larger and larger negative velocity.

3.21 Q: A cart travels with a constant nonzero acceleration along a straight line. Which graph best represents the relationship between the distance the cart travels and time of travel?

| (1) | (2) | (3) | (4) |

3.21 A: (1) A constant acceleration is caused by a linearly increasing velocity. Since velocity is obtained from the slope of the d-t graph, the d-t graph must be continually increasing to provide the correct v-t graph.

Kinematic Equations

Motion graphs such as the d-t, v-t, and a-t graphs are terrific tools for understanding motion. However, there are times when graphing motion may not be the most efficient or effective way of understanding the motion of an object. To assist in these situations, we add a set of problem-solving equations to our physics toolbox, known as the **kinematic equations**. The kinematic equations help us to solve for five key variables describing the motion of an object. Once you know the value of any three variables, you can use the kinematic equations to solve for the other two!

Variable	Meaning
v_i	initial velocity
v_f	final velocity
d	displacement
a	acceleration
t	time

$$v_f = v_i + at$$

$$v_f^2 = v_i^2 + 2ad$$

$$d = v_i t + \tfrac{1}{2}at^2$$

In using these equations to solve motion problems, it's important to take care in setting up your analysis before diving in to a solution. Key steps to solving kinematics problems include:

1. Labeling your analysis for horizontal (x-axis) or vertical (y-axis) motion.
2. Choosing and indicating a positive direction (typically the direction of initial motion).
3. Creating a motion analysis table (v_i, v_f, d, a, t).
4. Using what you know about the problem to fill in your "givens" in the table.
5. Once you know three items in the table, use kinematic equations to solve for any unknowns.
6. Verify that your solution makes sense.

Let's take a look at a sample problem to see how this strategy can be employed.

3.22 Q: A race car starting from rest accelerates uniformly at a rate of 4.90 meters per second². What is the car's speed after it has traveled 200 meters?

(1) 1960 m/s
(2) 62.6 m/s
(3) 44.3 m/s
(4) 31.3 m/s

3.22 A: Step 1: Horizontal Problem

Step 2: Positive direction is direction car starts moving.

Step 3 & 4:

Variable	Value
v_i	0 m/s
v_f	FIND
d	200 m
a	4.90 m/s²
t	?

Step 5: Choose a kinematic equation that includes the given information and the information sought, and solve for the unknown showing the initial formula, substitution with units, and answer with units (AFSA strategy).

$$v_f^2 = v_i^2 + 2ad$$

$$v_f^2 = (0\,^m\!/_s)^2 + 2(4.90\,^m\!/_{s^2})(200m)$$

$$v_f^2 = 1960\,^{m^2}\!/_{s^2}$$

$$v_f = \sqrt{1960\,^{m^2}\!/_{s^2}} = 44.3\,^m\!/_s$$

Step 6: (3) 44.3 m/s is one of the given answer choices, and is a reasonable speed for a race car (44.3 m/s is approximately 99 miles per hour).

This strategy works not only for horizontal motion problems, but also for vertical motion problems. Let's take a look at a "far-out" vertical motion problem.

3.23 Q: An astronaut standing on a platform on the Moon drops a hammer. If the hammer falls 6.0 meters vertically in 2.7 seconds, what is its acceleration?

(1) 1.6 m/s²

(2) 2.2 m/s²

(3) 4.4 m/s²

(4) 9.8 m/s²

3.23 A: Step 1: Vertical Problem

Step 2: Positive direction is down (direction hammer starts moving.)

Step 3 & 4: Note that a dropped object has an initial vertical velocity of 0 m/s.

Variable	Value
v_i	0 m/s
v_f	?
d	6 m
a	FIND
t	2.7 s

Step 5: Choose a kinematic equation that includes the given information and the information sought, and solve for the unknown showing the initial formula, substitution with units, and answer with units (AFSA strategy).

$$d = v_i t + \frac{1}{2} at^2 \xrightarrow{v_i = 0}$$

$$d = \frac{1}{2} at^2$$

$$a = \frac{2d}{t^2}$$

$$a = \frac{2(6m)}{(2.7s)^2} = 1.6 \, \text{m/s}^2$$

Step 6: (1) 1.6 m/s² is one of the given answer choices, and is less than the acceleration due to gravity on the surface of the Earth (9.81 m/s²). This answer can be verified further by searching on the Internet to confirm that the acceleration due to gravity on the surface of the moon is indeed 1.6 m/s².

In some cases, you may not be able to solve directly for the "find" quantity. In these cases, you can solve for the other unknown variable first, then choose an equation to give you your final answer.

3.24 Q: A car traveling on a straight road at 15.0 meters per second accelerates uniformly to a speed of 21.0 meters per second in 12.0 seconds. The total distance traveled by the car in this 12.0-second time interval is

(1) 36.0 m

(2) 180 m

(3) 216 m

(4) 252 m

3.24 A: Horizontal Problem, positive direction is forward.

Variable	Value
v_i	15 m/s
v_f	21 m/s
d	FIND
a	?
t	12 s

Can't find *d* directly, find *a* first.

$$v_f = v_i + at$$

$$a = \frac{v_f - v_i}{t}$$

$$a = \frac{21 \, \text{m/s} - 15 \, \text{m/s}}{12s} = 0.5 \, \text{m/s}^2$$

Now solve for d.

$$d = v_i t + \frac{1}{2}at^2$$

$$d = (15\,\text{m/s})(12s) + \frac{1}{2}(0.5\,\text{m/s}^2)(12s)^2$$

$$d = 216m$$

Check: (3) 216m is a given answer, and is reasonable, as this is greater than the 180m the car would have traveled if remaining at a constant speed of 15 m/s for the 12 second time interval.

3.25 Q: A car initially traveling at a speed of 16 meters per second accelerates uniformly to a speed of 20 meters per second over a distance of 36 meters. What is the magnitude of the car's acceleration?

(1) 0.11 m/s²

(2) 2.0 m/s²

(3) 0.22 m/s²

(4) 9.0 m/s²

3.25 A: (2) $v_f^2 = v_i^2 + 2ad$

$$a = \frac{v_f^2 - v_i^2}{2d}$$

$$a = \frac{(20\,\text{m/s})^2 - (16\,\text{m/s})^2}{2(36m)} = 2\,\text{m/s}^2$$

3.26 Q: An astronaut drops a hammer from 2.0 meters above the surface of the Moon. If the acceleration due to gravity on the Moon is 1.62 meters per second², how long will it take for the hammer to fall to the Moon's surface?

(1) 0.62 s

(2) 1.2 s

(3) 1.6 s

(4) 2.5 s

3.26 A: (3) $d = v_i t + \frac{1}{2}at^2 \xrightarrow{\;v_{i=0}\;}$

$$d = \frac{1}{2}at^2$$

$$t = \sqrt{\frac{2d}{a}} = \sqrt{\frac{2(2m)}{1.62\,\text{m/s}^2}} = 1.57s$$

3.27 Q: A car increases its speed from 9.6 meters per second to 11.2 meters per second in 4.0 seconds. The average acceleration of the car during this 4.0-second interval is

(1) 0.40 m/s²

(2) 2.4 m/s²

(3) 2.8 m/s²

(4) 5.2 m/s²

3.27 A: (1) $v_f = v_i + at$

$$a = \frac{v_f - v_i}{t} = \frac{(11.2\,{}^m/_s) - (9.6\,{}^m/_s)}{4s} = 0.4\,{}^m/_{s^2}$$

3.28 Q: A rock falls from rest a vertical distance of 0.72 meter to the surface of a planet in 0.63 seconds. The magnitude of the acceleration due to gravity on the planet is

(1) 1.1 m/s²

(2) 2.3 m/s²

(3) 3.6m/s²

(4) 9.8 m/s²

3.28 A: (3) $d = v_i t + \frac{1}{2}at^2 \xrightarrow{\;v_i=0\;}$

$$d = \frac{1}{2}at^2$$

$$a = \frac{2d}{t^2} = \frac{2(0.72m)}{(0.63s)^2} = 3.6\,{}^m/_{s^2}$$

3.29 Q: The speed of an object undergoing constant acceleration increases from 8.0 meters per second to 16.0 meters per second in 10 seconds. How far does the object travel during the 10 seconds?

(1) 3.6×10² m

(2) 1.6×10² m

(3) 1.2×10² m

(4) 8.0×10¹ m

3.29 A: Can't find d directly, find a first.

$$v_f = v_i + at$$

$$a = \frac{v_f - v_i}{t}$$

$$a = \frac{16\,{}^m/_s - 8\,{}^m/_s}{10s} = 0.8\,{}^m/_{s^2}$$

Now solve for d.

$$d = v_i t + \frac{1}{2}at^2$$

$$d = (8\,\tfrac{m}{s})(10s) + \tfrac{1}{2}(0.8\,\tfrac{m}{s^2})(10s)^2$$

$$d = 120m$$

Free Fall

Examination of free-falling bodies dates back to the days of Aristotle. At that time Aristotle believed that more massive objects would fall faster than less massive objects. He believed this in large part due to the fact that when examining a rock and a feather falling from the same height it is clear that the rock hits the ground first. Upon further examination it is clear that Aristotle was incorrect in his hypothesis.

As proof, take a basketball and a piece of paper. Drop them simultaneously from the same height... do they land at the same time? Probably not. Now take that piece of paper and crumple it up into a ball and repeat the experiment. Now what do you see happen? You should see that both the ball and the paper land at the same time. Therefore we can conclude that Aristotle's predictions did not account for the effect of air resistance. For the purposes of this course, we will neglect drag forces such as air resistance.

In the 17th century, Galileo Galilei began a re-examination of the motion of falling bodies. Galileo, recognizing that air resistance affects the motion of a falling body, executed his famous thought experiment in which he continuously asked what would happen if the effect of air resistance was removed. Commander David Scott of Apollo 15 performed this experiment while on the moon. He simultaneously dropped a hammer and a feather, and observed that they reached the ground at the same time.

Since Galileo's experiments, scientists have come to a better understanding of how the gravitational pull of the Earth accelerates free-falling bodies. Through experimentation it has been determined that on the surface of the Earth all objects experience acceleration (g) of 9.81 m/s² toward the center of the Earth. (NOTE: If we move off the surface of the Earth the acceleration due to gravity changes.)

We can look at free-falling bodies as objects being dropped from some height or thrown vertically upward. In this examination we will analyze the motion of each condition.

Objects Falling From Rest

Objects starting from rest have an initial velocity of zero, giving us our first kinematic quantity we'll need for problem solving. Beyond that, if we call the direction of initial motion (down) positive, the object will have a positive acceleration and speed up as it falls.

An important first step in analyzing objects in free fall is deciding which direction along the y-axis we are going to call positive and which direction will therefore be negative. Although you can set your positive direction any way you want and get the correct answer, following the hints below can simplify your work to reach the correct answer consistently.

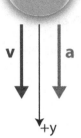

1. Identify the direction of the object's initial motion and assign that as the positive direction. In the case of a dropped object, the positive y-direction will point toward the bottom of the paper.
2. With the axis identified you can now identify and write down your given kinematic information. Don't forget that a dropped object has an initial velocity of zero.
3. Notice the direction the vector arrows are drawn — if the velocity and acceleration point in the same direction, the object speeds up. If they point in opposite directions, the object slows down.

3.30 Q: What is the speed of a 2.5-kilogram mass after it has fallen freely from rest through a distance of 12 meters?
  (1) 4.8 m/s
  (2) 15 m/s
  (3) 30 m/s
  (4) 43 m/s

3.30 A: Vertical Problem: Declare down as the positive direction. This means that the acceleration, which is also down, is a positive quantity.

Variable	Value
v_i	0 m/s
v_f	FIND
d	12 m
a	9.81 m/s²
t	?

$$v_f^2 = v_i^2 + 2ad$$
$$v_f^2 = (0\,{}^m\!/_s)^2 + 2(9.81\,{}^m\!/_{s^2})(12m)$$
$$v_f^2 = 235\,{}^{m^2}\!/_{s^2}$$
$$v_f = \sqrt{235\,{}^{m^2}\!/_{s^2}} = 15.3\,{}^m\!/_s$$

Correct answer is (2) 15 m/s.

3.31 Q: How far will a brick starting from rest fall freely in 3.0 seconds?

    (1) 15 m

    (2) 29 m

    (3) 44 m

    (4) 88 m

3.31 A: (3) 44m

Variable	Value
v_i	0 m/s
v_f	?
d	FIND
a	9.81 m/s²
t	3 s

$$d = v_i t + \tfrac{1}{2} at^2$$

$$d = (0 \, ^m\!/_s)(3s) + \tfrac{1}{2}(9.81 \, ^m\!/_s)(3s)^2$$

$$d = 44m$$

3.32 Q: A ball dropped from rest falls freely until it hits the ground with a speed of 20 meters per second. The time during which the ball is in free fall is approximately

    (1) 1 s

    (2) 2 s

    (3) 0.5 s

    (4) 10 s

3.32 A: (2) 2 s

Variable	Value
v_i	0 m/s
v_f	20 m/s
d	?
a	9.81 m/s²
t	FIND

$$v_f = v_i + at$$

$$t = \frac{v_f - v_i}{a}$$

$$t = \frac{20 \, ^m\!/_s - 0 \, ^m\!/_s}{9.81 \, ^m\!/_{s^2}} = 2.04s$$

Objects Launched Upward

Examining the motion of an object being launched vertically upward is done in much the same way we examined the motion of an object falling from rest. The major difference is that we have to look at two components to its motion instead of one: both up *and* down.

Before we get into establishing our frame of reference and working through the quantitative analysis let's build a solid conceptual understanding of what is happening while the ball is in the air. Consider the ball being thrown vertically into the air as shown in the diagram.

In order for the ball to move upwards its initial velocity must be greater than zero. As the ball rises, its velocity decreases until it reaches its maximum height, stops, and begins to fall. As the ball falls, its speed increases. In other words, the ball is accelerating the entire time it is in the air, both on the way up, at the instant it stops at its highest point, and on the way down.

The cause of the ball's acceleration is gravity. The entire time the ball is in the air, its acceleration is 9.81 m/s² provided this occurs on the surface of the Earth. Note that the acceleration can be either 9.81 m/s² or -9.81 m/s². The sign of the acceleration depends on the direction we declared as positive, but in all cases the direction of the acceleration due to gravity is down, toward the center of the Earth.

We have already shown the ball's acceleration for the entire time it is in the air is 9.81m/s² down. This acceleration causes the ball's velocity to decrease at a constant rate until it reaches maximum altitude, at which point it turns around and starts to fall. In order to turn around the ball's velocity must pass through zero. Therefore, at maximum altitude the velocity of the ball must be zero.

3.33 Q: A ball thrown vertically upward reaches a maximum height of 30 meters above the surface of Earth. At its maximum height, the speed of the ball is

(1) 0 m/s

(2) 3.1 m/s

(3) 9.8 m/s

(4) 24 m/s

3.33 A: (1) 0 m/s. The instantaneous speed of any projectile at its maximum height is zero.

Because gravity provides the same acceleration to the ball on the way up (slowing it down) as on the way down (speeding it up), the time to reach maximum altitude is the same as the time to return to its launch position. In similar fashion, the initial velocity of the ball on the way up will equal the velocity of the ball at the instant it reaches the point from which it was launched on the way down. Put another way, the time to go up is equal to the time to go down, and the initial velocity up is equal to the final velocity down (assuming the object begins and ends at the same height above ground).

Now that a conceptual understanding of the ball's motion has been established, we can work toward a quantitative solution. As with any problem, that solution starts with establishing our frame of reference. Following the rule of thumb established previously, we will assign the direction the ball begins to move as positive.

Remember that assigning positive and negative directions are completely arbitrary. You have the freedom to assign them how you see fit. Once you assign them, however, don't change them.

Once this positive reference direction has been established, all other velocities and displacements are assigned accordingly. For example, if up is the positive direction, the acceleration due to gravity will be negative, because the acceleration due to gravity points down, toward the center of the Earth. At its highest point, the ball will have a positive displacement, and will have a zero displacement when it returns to its starting point. If the ball isn't caught, but continues toward the Earth past its starting point, it will have a negative displacement.

A "trick of the trade" to solving free fall problems involves symmetry. The time an object takes to reach its highest point is equal to the time it takes to return to the same vertical position. The speed with which the projectile begins its journey upward is equal to the speed of the projectile when it returns to the same height (although, of course, its velocity is in the opposite direction). If you want to simplify the problem, vertically, at its highest point, the vertical velocity is 0. This added information can assist you in filling out your vertical motion tables. If you cut the object's motion in half, you can simplify your problem solving – but don't forget that if you want the total time in the air, you must double the time it takes for the object to rise to its highest point.

3.34 Q: A basketball player jumped straight up to grab a rebound. If she was in the air for 0.80 seconds, how high did she jump?

    (1) 0.50 m

    (2) 0.78 m

    (3) 1.2 m

    (4) 3.1 m

3.34 A: Define up as the positive y-direction. Note that if basketball player is in the air for 0.80 seconds, she reaches her maximum height at a time of 0.40 seconds, at which point her velocity is zero.

Variable	Value
v_i	?
v_f	0 m/s
d	FIND
a	-9.81 m/s²
t	0.40 s

Can't solve for d directly with given information, so find v_i first.

$$v_f = v_i + at$$

$$v_i = v_f - at$$

$$v_i = 0 - (-9.81 \, ^m/_{s^2})(0.40s) = 3.92 \, ^m/_s$$

Now with v_i known, solve for d.

$$d = v_i t + \tfrac{1}{2}at^2$$

$$d = (3.92 \, ^m/_s)(0.40s) + \tfrac{1}{2}(-9.81 \, ^m/_{s^2})(0.40s)^2$$

$$d = 0.78m$$

Correct answer is (2) 0.78 m. This is a reasonable height for a basketball player to jump.

3.35 Q: Which graph best represents the relationship between the acceleration of an object falling freely near the surface of Earth and the time that it falls?

      (1)   (2)   (3)   (4)

3.35 A: (4) The acceleration due to gravity is a constant 9.81 m/s² down on the surface of the Earth.

3.36 Q: A ball is thrown straight downward with a speed of 0.50 meter per second from a height of 4.0 meters. What is the speed of the ball 0.70 seconds after it is released?

(1) 0.50 m/s

(2) 7.4 m/s

(3) 9.8 m/s

(4) 15 m/s

3.36 A: (2) 7.4 m/s. Note that in filling out the kinematics table, the height of 4 meters is not the displacement of the ball, but is extra unneeded information.

Variable	Value
v_i	0.50 m/s
v_f	FIND
d	?
a	9.81 m/s²
t	0.70 s

$$v_f = v_i + at$$

$$v_f = 0.50\,{}^m\!/_s + (9.81\,{}^m\!/_{s^2})(0.70s)$$

$$v_f = 7.4\,{}^m\!/_s$$

3.37 Q: A baseball dropped from the roof of a tall building takes 3.1 seconds to hit the ground. How tall is the building? [Neglect friction.]

(1) 15 m

(2) 30 m

(3) 47 m

(4) 94 m

3.37 A: (3) 47 m

$$d = v_i t + \tfrac{1}{2}at^2$$

$$d = (0\,{}^m\!/_s)(3.1s) + \tfrac{1}{2}(9.81\,{}^m\!/_{s^2})(3.1s)^2$$

$$d = 47m$$

3.38 Q: A 0.25-kilogram baseball is thrown upward with a speed of 30 meters per second. Neglecting friction, the maximum height reached by the baseball is approximately

(1) 15 m

(2) 46 m

(3) 74 m

(4) 92 m

3.38 A: (2) 46m

$$v_f^2 = v_i^2 + 2ad$$

$$d = \frac{v_f^2 - v_i^2}{2a}$$

$$d = \frac{(0\,{}^m\!/_s)^2 - (30\,{}^m\!/_s)^2}{2(-9.81\,{}^m\!/_{s^2})} = 45.9m$$

Projectile Motion

Projectile motion problems, or problems of an object launched in both the x- and y- directions, can be analyzed using the physics you already know. The key to solving these types of problems is realizing that the horizontal component of the object's motion is independent of the vertical component of the object's motion. Since you already know how to solve horizontal and verti- cal kinematics problems, all you have to do is put the two results together!

Start these problems by making separate motion tables for vertical and hori- zontal motion. Vertically, the setup is the same for projectile motion as it is for an object in free fall. Horizontally, gravity only pulls an object down, it never pulls or pushes an object horizontally, therefore the horizontal accelera- tion of any projectile is zero. If the acceleration horizontally is zero, velocity must be constant, therefore v_i horizontally must equal v_f horizontally. Finally, to tie the problem together, realize that the time the projectile is in the air vertically must be equal to the time the projectile is in the air horizontally.

When an object is launched or thrown completely horizontally, such as a rock thrown horizontally off a cliff, the initial velocity of the object is its initial horizontal velocity. Because horizontal velocity doesn't change, this velocity is also the object's final horizontal velocity, as well as its average horizontal velocity. Further, the initial vertical velocity of the projectile is zero. This means that you could hurl an object 1000 m/s horizontally off a cliff, and si- multaneously drop an object off the cliff from the same height, and they will both reach the ground at the same time (even though the hurled object has traveled a greater distance).

3.39Q: Fred throws a baseball 42 m/s horizontally from a height of 2m. How far will the ball travel before it reaches the ground?

3.39A: To solve this problem, we must first find how long the ball will remain in the air. This is a vertical motion problem.

VERTICAL MOTION TABLE

Variable	Value
v_i	0 m/s
v_f	?
d	2 m
a	9.81 m/s²
t	FIND

$$d = v_i t + \tfrac{1}{2} a t^2$$

$$d = \tfrac{1}{2} a t^2$$

$$t = \sqrt{\frac{2d}{a}}$$

$$t = \sqrt{\frac{2(2m)}{9.81 \; ^m\!/_s}} = 0.639s$$

Now that we know the ball is in the air for 0.639 seconds, we can find how far it travels horizontally before reaching the ground. This is a horizontal motion problem, in which the acceleration is 0 (nothing is causing the ball to accelerate horizontally.) Because the ball doesn't accelerate, its initial velocity is also its final velocity, which is equal to its average velocity.

HORIZONTAL MOTION TABLE

Variable	Value
v_i	42 m/s
v_f	42 m/s
d	FIND
a	0 m/s²
t	0.639 s

$$\overline{v} = \frac{d}{t}$$

$$d = \overline{v}t = (42 \; ^m\!/_s)(0.639s) = 26.8m$$

We can therefore conclude that the baseball travels 26.8 meters horizontally before reaching the ground.

3.40 Q: The diagram below represents the path of a stunt car that is driven off a cliff, neglecting friction.

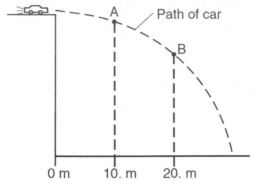

Compared to the horizontal component of the car's velocity at point A, the horizontal component of the car's velocity at B is

(1) smaller

(2) greater

(3) the same

3.40 A: (3) the same. The car's horizontal acceleration is zero, therefore the horizontal velocity remains constant.

3.41 Q: A 0.2-kilogram red ball is thrown horizontally at a speed of 4 meters per second from a height of 3 meters. A 0.4-kilogram green ball is thrown horizontally from the same height at a speed of 8 meters per second. Compared to the time it takes the red ball to reach the ground, the time it takes the green ball to reach the ground is

(1) one-half as great

(2) twice as great

(3) the same

(4) four times as great

3.41 A: (3) the same. Both objects are thrown horizontally from the same height. Because horizontal motion and vertical motion are independent, both objects have the same vertical motion (they both start with an initial vertical velocity of 0, have the same acceleration of 9.81 m/s^2 down, and both travel the same vertical distance). Therefore, the two objects reach the ground in the same amount of time.

3.42 Q: A ball is thrown horizontally at a speed of 24 meters per second from the top of a cliff. If the ball hits the ground 4.0 seconds later, approximately how high is the cliff?

(1) 6.0 m

(2) 39 m

(3) 78 m

(4) 96 m

3.42 A: (3) 78m

$$d = v_i t + \tfrac{1}{2} a t^2$$
$$d = \tfrac{1}{2} a t^2$$
$$d = \tfrac{1}{2}(9.81 \, ^m\!/_{s^2})(4s)^2$$
$$d = 78m$$

3.43 Q: Projectile A is launched horizontally at a speed of 20 meters per second from the top of a cliff and strikes a level surface below, 3.0 seconds later. Projectile B is launched horizontally from the same location at a speed of 30 meters per second. The time it takes projectile B to reach the level surface is

(1) 4.5 s

(2) 2.0 s

(3) 3.0 s

(4) 10 s

3.43 A: (3) 3.0 s. They both take the same time to reach the ground because they both travel the same distance vertically, and they both have the same vertical acceleration (9.81 m/s² down) and initial vertical velocity (zero).

Angled Projectiles

For objects launched at an angle, you have to do a little more work to de-termine the initial velocity in both the horizontal and vertical directions. For example, if a football is kicked with an initial velocity of 40 m/s at an angle of 30° above the horizontal, you need to break the initial velocity vector up into x- and y-components in the same manner as covered in the components of vectors math review section.

Then, use the components for your initial velocities in your horizontal and vertical tables. Finally, don't forget that symmetry of motion also applies to the parabola of projectile motion. For objects launched and landing at the same height, the launch angle is equal to the landing angle. The launch velocity is equal to the landing velocity. And if you want an object to travel the maximum possible horizontal distance (or range), launch it at an angle of 45°.

3.44 Q: Herman the human cannonball is launched from level ground at an angle of 30° above the horizontal with an initial velocity of 26 m/s. How far does Herman travel horizontally before reuniting with the ground?

3.44 A: Our first step in solving this type of problem is to determine Herman's initial horizontal and vertical velocity. We do this by breaking up his initial velocity into vertical and horizontal components:

$$v_{i_x} = v_i \cos(\theta) = (26\,{}^m\!/\!_s)\cos(30°) = 22.5\,{}^m\!/\!_s$$

$$v_{i_y} = v_i \sin(\theta) = (26\,{}^m\!/\!_s)\sin(30°) = 13\,{}^m\!/\!_s$$

Next, we'll analyze Herman's vertical motion to find out how long he is in the air. We'll analyze his motion on the way up, find the time, and double that to find his total time in the air:

VERTICAL MOTION TABLE

Variable	Value
v_i	13 m/s
v_f	0 m/s
d	?
a	-9.81 m/s²
t	FIND

$$v_f = v_i + at$$

$$t = \frac{v_f - v_i}{a}$$

$$t_{up} = \frac{(0 - 13\,{}^m\!/\!_s)}{-9.81\,{}^m\!/\!_{s^2}} = 1.33s$$

$$t_{total} = 2 \times t_{up} = 2.65s$$

Now that we know Herman was in the air 2.65s, we can find how far he moved horizontally, using his initial horizontal velocity of 22.5 m/s.

HORIZONTAL MOTION TABLE

Variable	Value
v_i	22.5 m/s
v_f	22.5 m/s
d	FIND
a	0
t	2.65s

$$\overline{v} = \frac{d}{t}$$

$$d = \overline{v}t$$

$$d = (22.5\,{}^m\!/\!_s)(2.65s) = 59.6m$$

Therefore, Herman must have traveled 59.6m horizontally before returning to the Earth.

3.45 Q: A child kicks a ball with an initial velocity of 8.5 meters per second at an angle of 35º with the horizontal, as shown. The ball has an initial vertical velocity of 4.9 meters per second and a total time of flight of 1.0 second. The maximum height reached by the ball is approximately: [Neglect air resistance]

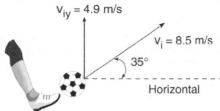

(1) 1.2 m

(2) 2.5 m

(3) 4.9 m

(4) 8.5 m

3.45 A: The maximum height in the air is a vertical motion problem. Start by recognizing that at its maximum height, the ball's vertical velocity will be zero, and it will have been in the air 0.5 seconds.

VERTICAL MOTION TABLE

Variable	Value
v_i	4.9 m/s
v_f	0 m/s
d	FIND
a	-9.81 m/s²
t	0.5 s

$$v_f^2 = v_i^2 + 2ad$$

$$d = \frac{v_f^2 - v_i^2}{2a}$$

$$d = \frac{(0\,^m/_s)^2 - (4.9\,^m/_s)^2}{2(-9.81\,^m/_{s^2})} = 1.2m$$

The correct answer must be (1) 1.2 meters. Note that you could have solved for the correct answer using any of the kinematic equations containing distance.

3.46 Q: A ball is thrown at an angle of 38° to the horizontal. What happens to the magnitude of the ball's vertical acceleration during the total time interval that the ball is in the air?

(1) It decreases, then increases.

(2) It decreases, then remains the same.

(3) It increases, then decreases.

(4) It remains the same.

3.46 A: (4) It remains the same since the acceleration of any projectile on the surface of Earth is 9.81 m/s^2 down the entire time the projectile is in the air.

3.47 Q: A golf ball is hit at an angle of 45° above the horizontal. What is the acceleration of the golf ball at the highest point in its trajectory? [Neglect friction.]

(1) 9.8 m/s^2 upward

(2) 9.8 m/s^2 downward

(3) 6.9 m/s^2 horizontal

(4) 0 m/s^2.

3.47 A: (2) 9.8 m/s^2 downward.

3.48 Q: A machine launches a tennis ball at an angle of 25° above the horizontal at a speed of 14 meters per second. The ball returns to level ground. Which combination of changes must produce an increase in time of flight of a second launch?

(1) decrease the launch angle and decrease the ball's initial speed

(2) decrease the launch angle and increase the ball's initial speed

(3) increase the launch angle and decrease the ball's initial speed

(4) increase the launch angle and increase the ball's initial speed

3.48 A: (4) will increase the ball's initial vertical velocity and therefore give the ball a larger time of flight.

3.49 Q: A golf ball is given an initial speed of 20 meters per second and returns to level ground. Which launch angle above level ground results in the ball traveling the greatest horizontal distance? [Neglect friction.]

(1) 60°

(2) 45°

(3) 30°

(4) 15°

3.49 A: (2) 45° provides the greatest range for a projectile launched from level ground onto level ground, neglecting friction.

You can find more practice problems on the APlusPhysics website at: http://www.aplusphysics.com/regents.

Chapter 4: Dynamics

"If I have seen further than others, it is by standing upon the shoulders of giants."

— Sir Isaac Newton

Objectives

1. Define mass and inertia and explain the meaning of Newton's 1st Law.
2. Define a force and distinguish between contact forces and field forces.
3. Draw and label a free body diagram showing all forces acting on an object.
4. Determine the resultant of two or more vectors graphically and algebraically.
5. Draw scaled force diagram using a ruler and protractor.
6. Resolve a vector into perpendicular components: both graphically and algebraically.
7. Use vector diagrams to analyze mechanical systems (equilibrium and nonequilibrium).
8. Verify Newton's Second Law for linear motion.
9. Describe how mass and weight are related.
10. Define friction and distinguish between static and kinetic friction.
11. Determine the coefficient of friction for two surfaces.
12. Calculate parallel and perpendicular components of an object's weight to solve ramp problems.

Now that we've studied kinematics, you should have a pretty good understanding that objects in motion have kinetic energy, which is the ability of a moving object to move another object. To change an object's motion, and therefore its kinetic energy, the object must undergo a change in velocity, which is called an acceleration. So then, what causes an acceleration? To answer that question, we must study forces and their application.

Dynamics, or the study of forces, was very simply and effectively described by Sir Isaac Newton in 1686 in his masterpiece <u>Principia Mathematica Philosophiae Naturalis</u>. Newton described the relationship between forces and motion using three basic principles. Known as Newton's Laws of Motion, these concepts are still used today in applications ranging from sports science to aeronautical engineering.

Newton's 1st Law of Motion

Newton's 1st Law of Motion, also known as the **law of inertia**, can be summarized as follows:

> "An object at rest will remain at rest, and an object in motion will remain in motion, at constant velocity and in a straight line, unless acted upon by a net force."

This means that unless there is a net (unbalanced) force on an object, an object will continue in its current state of motion with a constant velocity. If this velocity is zero (the object is at rest), the object will continue to remain at rest. If this velocity is not zero, the object will continue to move in a straight line at the same speed. However, if a net (unbalanced) force does act on an object, that object's velocity will be changed (it will accelerate).

This sounds like a simple concept, but it can be quite confusing because it is difficult to observe this in everyday life. We're usually fine with understanding the first part of the law: "an object at rest will remain at rest unless acted upon by a net force." This is easily observable. The donut sitting on your breakfast table this morning didn't spontaneously accelerate up into the sky. Nor did the family cat, Whiskers, lounging sleepily on the couch cushion the previous evening, all of a sudden accelerate sideways off the couch for no apparent reason.

The second part of the law contributes a considerably bigger challenge to our conceptual understanding. Realizing that "an object in motion will continue in its current state of motion with constant velocity unless acted upon by a net force" isn't easy to observe here on Earth, making this law rather tricky. Almost all objects we observe in our everyday lives that are in motion are being acted upon by a net force - friction. Try this example: take your physics book and give it a good push along the floor. As expected, the book moves for some distance, but rather rapidly slides to a halt. An outside force, friction, has acted upon it. Therefore, from our typical observations, it would be easy to think that an object must have a force continually applied upon it to remain in motion. However, this isn't so. If we took the same book out into the far reaches of space, away from any gravitational or frictional forces, and pushed it away from us, it would continue moving in a straight line at a constant velocity forever and ever, as there are no external forces to change its motion. When the net force on an object is 0, we say the object is in static equilibrium. We'll revisit static equilibrium when we talk about Newton's 2nd Law.

The tendency of an object to resist a change in velocity is known as the object's **inertia**. For example, a train has significantly more inertia than a skateboard. It is much harder to change the train's velocity than it is the skateboard's. The measure of an object's inertia is its mass. For the purposes of this course, inertia and mass mean the same thing - they are synonymous.

4.1 Q: A 0.50-kilogram cart is rolling at a speed of 0.40 meter per second. If the speed of the cart is doubled, the inertia of the cart is

(1) halved
(2) doubled
(3) quadrupled
(4) unchanged

4.1 A: (4) unchanged. Inertia is another word for mass, and the mass of the cart is constant.

4.2 Q: Which object has the greatest inertia?
(1) a falling leaf
(2) a softball in flight
(3) a seated high school student
(4) a rising helium-filled toy balloon

4.2 A: (3) a seated high school student has the greatest inertia (mass).

4.3 Q: Which object has the greatest inertia?
(1) a 5.00-kg mass moving at 10.0 m/s
(2) a 10.0-kg mass moving at 1.00 m/s
(3) a 15.0-kg mass moving at 10.0 m/s
(4) a 20.0-kg mass moving at 1.00 m/s

4.3 A: (4) a 20.0-kg mass has the greatest inertia.

If you recall from the kinematics unit, a change in velocity is known as an acceleration. Therefore, the second part of this law could be re-written to state that an object acted upon by a net force will be accelerated. A more detailed description of this acceleration is given by Newton's 2nd Law.

But, what exactly is a force? A **force** is a vector quantity describing the push or a pull on an object. Forces are measured in Newtons (N), named after Sir Isaac Newton, of course. A Newton is not a base unit, but is instead a derived unit, equivalent to 1 kg$\times$m/s^2. Interestingly, the gravitational force on a medium-sized apple is approximately 1 Newton.

We can break forces down into two basic types: contact forces and field forces. Contact forces occur when objects touch each other. Examples of contact forces include pushing a crate (applied force), pulling a wagon (tension force), a frictional force slowing down your sled, or even the force of air accelerating a spitwad through a straw. Field forces, also known as non-contact forces, occur at a distance. Examples of field forces include the gravitational force, the magnetic force, and the electrical force between two charged objects.

So, what then is a net force? A **net force** is just the vector sum of all the forces acting on an object. Imagine you and your sister are fighting over the last Christmas gift. You are pulling one end of the gift toward you with a force of 5N. Your sister is pulling the other end toward her (in the opposite direction) with a force of 5N. The net force on the gift, then, would be 0N, therefore there would be no net force. As it turns out, though, you have a passion for Christmas gifts, and now increase your pulling force to 6N. The net force on the gift now is 1N in your direction, therefore the gift would begin to accelerate toward you (yippee!) It can be difficult to keep track of all the forces acting on an object.

Free Body Diagrams

Fortunately, we have a terrific tool for helping us to analyze the forces acting upon objects. This tool is known as a free body diagram. Quite simply, a **free body diagram** is just a representation of a single object, or system, with vector arrows showing all the external forces acting on the object. These diagrams make it very easy to identify exactly what the net force is on an object, and they're also quite simple to create:

1. Isolate the object of interest. Draw the object as a point par-

ticle representing the same mass.

2. Sketch and label each of the external forces acting on the object.
3. Choose a coordinate system, with the direction of motion as one of the positive coordinate axes.
4. If all forces do not line up with your axes, resolve those forces into components using trigonometry.

$$A_x = A\cos(\theta)$$

$$A_y = A\sin(\theta)$$

5. Redraw your free body diagram, replacing forces that don't overlap the coordinates axes with their components.

As an example, let's picture a glass of soda sitting on the dining room table. We can represent our object (the glass of soda) in the diagram as a single dot. Then, we represent each of the vector forces acting on the soda by drawing arrows and labeling them. In this case, we can start by recognizing the force of gravity on the soda, known more commonly as the soda's **weight**. Although we could label this force as F_{grav}, or W, we'll get in the habit right now of writing the force of gravity on an object as mg. We can do this because the force of gravity on an object is equal to the object's mass times the acceleration due to gravity, g.

Of course, since the soda isn't accelerating, there must be another force acting on the soda to balance out the weight. This force, the force of the table pushing up on the soda, is known as the **normal force (F_N)**. In physics, the normal force refers to a force perpendicular to a surface (normal in this case meaning perpendicular). The force of gravity on the soda must exactly match the normal force on the soda, although they are in opposite directions, therefore there is no net force on the soda. The free body diagram for this situation could be drawn as shown at left.

4.4 Q: Which diagram represents a box in equilibrium?

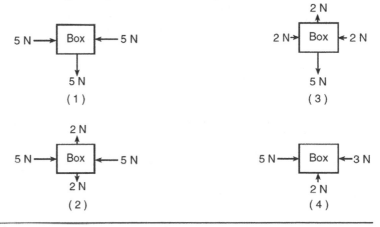

4.4 A: (2) all forces are balanced for a net force of zero.

4.5 Q: If the sum of all the forces acting on a moving object is zero, the object will
(1) slow down and stop
(2) change the direction of its motion
(3) accelerate uniformly
(4) continue moving with constant velocity

4.5 A: (4) continue moving with constant velocity per Newton's 1st Law.

Newton's 2nd Law of Motion

Newton's 2nd Law of Motion may be the most important principle in all of modern-day physics. It explains exactly how an object's velocity is changed by a net force. In words, Newton's 2nd Law states that "the acceleration of an object is directly proportional to the net force applied, and inversely proportional to the object's mass." In equation form:

$$a = \frac{F_{net}}{m}$$

It's important to remember that both force and acceleration are vectors. Therefore, the direction of the acceleration, or the change in velocity, will be in the same direction as the net force. You can also look at this equation from the opposite perspective. A net force applied to an object changes an object's velocity (produces an acceleration), and is frequently written as:

$$F_{net} = ma$$

You can analyze many situations involving both balanced and unbalanced forces on an object using the same basic steps.

1. Draw a free body diagram.
2. For any forces that don't line up with the x- or y-axes, break those forces up into components that do lie on the x- or y-axis.
3. Write expressions for the net force in x- and y- directions. Set the net force equal to ma, since Newton's 2nd Law tells us that F=ma.
4. Solve the resulting equations.

Let's take a look and see how these steps can be applied to a sample problem.

4.6 Q: A force of 25 newtons east and a force of 25 newtons west act concurrently on a 5-kilogram cart. Find the acceleration of the cart.

(1) 1.0 m/s² west

(2) 0.20 m/s² east

(3) 5.0 m/s² east

(4) 0 m/s²

4.6 A: Step 1: Draw a free-body diagram (FBD).

$$25\,\text{N} \qquad 25\text{N}$$

Step 2: All forces line up with x-axis. Define east as positive.

Step 3: $F_{net} = 25N - 25N = ma$

Step 4: $0 = ma$

$a - 0$

Correct answer must be (4) 0 m/s².

Of course, everything we've already learned about kinematics still applies, and can be applied to dynamics problems as well.

4.7 Q: A 0.15-kilogram baseball moving at 20 m/s is stopped by a catcher in 0.010 seconds. The average force stopping the ball is

(1) 3.0×10^{-2} N

(2) 3.0×10^{0} N

(3) 3.0×10^{1} N

(4) 3.0×10^{2} N

4.7 A: First write down what information is given and what we're asked to find. Define the initial direction of the baseball as positive.

Given: Find:

$m = 0.15kg$ F

$v_i = 20\,{}^{m}\!/_{s}$

$v_f = 0\,{}^{m}\!/_{s}$

$t = 0.010s$

Use kinematics to find acceleration.

$$a = \frac{\Delta v}{t} = \frac{v_f - v_i}{t} \Rightarrow$$

$$a = \frac{0\,{}^{m}\!/_{s} - 20\,{}^{m}\!/_{s}}{0.010s} = -2000\,{}^{m}\!/_{s^2}$$

The negative acceleration indicates the acceleration is in the direction opposite that of the initial velocity of the baseball. Now that we know acceleration, we can solve for force using Newton's 2nd Law.

$$F_{net} = ma = (0.15)(-2000 \, ^m\!/\!_{s^2}) = -300N$$

The correct answer must be (4), 300 newtons. The negative sign in our answer indicates that the force applied is opposite the direction of the baseball's initial velocity.

4.8 Q: Two forces, F_1 and F_2, are applied to a block on a frictionless, horizontal surface as shown below.

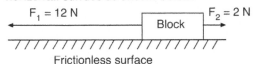

If the magnitude of the block's acceleration is 2.0 meters per second², what is the mass of the block?

(1) 1 kg

(2) 5 kg

(3) 6 kg

(4) 7 kg

4.8 A: Define left as the positive direction.

$$F_{net} = 12N - 2N = 10N = ma$$

$$m = \frac{F_{net}}{a} = \frac{10N}{2 \, ^m\!/\!_{s^2}} = 5kg$$

(2) 5 kg

4.9 Q: What is the weight of a 2.00-kilogram object on the surface of Earth?

(1) 4.91 N

(2) 2.00 N

(3) 9.81 N

(4) 19.6 N

4.9 A: Weight is the force of gravity on an object. From Newton's 2nd Law, the Force of gravity on an object (F_g), is equal to the mass of the object times its acceleration, the acceleration due to gravity (9.81 m/s²), which we abbreviate as g.

$$F_g = ma$$

$$W = mg$$

$$W = (2kg)(9.81 \, {}^m\!/\!{}_{s^2}) = 19.6N$$

(4) 19.6 N is correct.

4.10 Q: A 25-newton horizontal force northward and a 35-newton horizontal force southward act concurrently on a 15-kilogram object on a frictionless surface. What is the magnitude of the object's acceleration?

(1) 0.67 m/s²

(2) 1.7 m/s²

(3) 2.3 m/s²

(4) 4.0 m/s²

4.10 A: (1) 0.67 m/s².

$$a = \frac{F_{net}}{m}$$

$$a = \frac{35N - 25N}{15kg} = 0.67 \, {}^m\!/\!{}_{s^2}$$

Static Equilibrium

The special situation in which the net force on an object turns out to be zero, which we call **static equilibrium**, tells us immediately that the object isn't accelerating. If the object is moving with some velocity, it will remain moving with that exact same velocity. If the object is at rest, it will remain at rest. Sounds familiar, doesn't it? This is a restatement of Newton's 1st Law of Motion, the Law of Inertia. So in reality, Newton's 1st Law is just a special case of Newton's 2nd Law, describing static equilibrium conditions! Consider the situation of a tug-of-war... if both participants are pulling with tremendous force, but the force is balanced, there is no acceleration -- a great example of static equilibrium.

Static equilibrium conditions are so widespread that knowing how to explore and analyze these conditions is a key stepping stone to understanding more complex situations.

One common analysis question involves finding the equilibrant force given a free body diagram of an object. The
equilibrant is a single force vector that you add to the unbalanced forces on an object in order to bring it into static
equilibrium. For example, if you are given a force vector of
10N north and 10N east, and asked to find the equilibrant,
you're really being asked to find a force that will offset the
two given forces, bringing the object into static equilibrium.

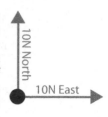

To find the equilibrant, you must first find the net force
being applied to the object. To do this, we revisit our vector math and add up the two vectors by first lining them
up tip to tail, then drawing a straight line from the starting point of the first vector to the ending point of the last
vector. The magnitude of this vector can be found from
the Pythagorean Theorem.

Finally, to find the equilibrant vector, we need to add a single vector
to the diagram that will give a net force of zero. If our total net
force is currently 14N northeast, then the vector that should
bring this back into equilibrium, the equilibrant, must be
the opposite of 14N northeast, or a vector with magnitude
14N to the southwest.

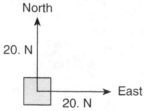

4.11 Q: A 20-newton force due north and a 20-newton force due east act
concurrently on an object, as shown in the diagram below.

North

20. N

East

20. N

The additional force necessary to bring the object into a state of
equilibrium is

(1) 20 N northeast

(2) 20 N southwest

(3) 28 N northeast

(4) 28 N southwest

4.11 A: (4) The resultant vector is 28 newtons northeast, so its equilibrant
must be 28 newtons southwest

Another common analysis question involves asking whether three vectors could be arranged to provide a static equilibrium situation.

4.12 Q: A 3-newton force and a 4-newton force are acting concurrently on a point. Which force could not produce equilibrium with these two forces?

(1) 1 N

(2) 7 N

(3) 9 N

(4) 4 N

4.12 A: (3) A 9-newton force could not produce equilibrium with a 3-newton and a 4-newton force.

4.13 Q: A net force of 10 newtons accelerates an object at 5.0 meters per second². What net force would be required to accelerate the same object at 1.0 meter per second²?

(1) 1.0 N

(2) 2.0 N

(3) 5.0 N

(4) 50 N

4.13 A: Strategy: First, solve for the mass of the object.

$$F_{net} = ma$$

$$m = \frac{F_{net}}{a} = \frac{10N}{5\,m/_{s^2}} = 2kg$$

Next, use Newton's 2nd Law to determine the force required to accelerate the object at 1 m/s².

$$F_{net} = ma = (2kg)(1\,m/_{s^2}) = 2N$$

The correct answer is (2) 2.0 N.

4.14 Q: A 1.0-newton metal disk rests on an index card that is balanced on top of a glass. What is the net force acting on the disk?

(1) 1 N

(2) 2 N

(3) 0 N

(4) 9.8 N

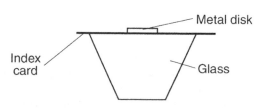

4.14 A: (3) 0N because the disk is at rest, so the net force must be zero.

4.15 Q: A 1200-kilogram space vehicle travels at 4.8 meters per second along the level surface of Mars. If the magnitude of the gravitational field strength on the surface of Mars is 3.7 newtons per kilogram, the magnitude of the normal force acting on the vehicle is

(1) 320 N

(2) 930 N

(3) 4400 N

(4) 5800 N

4.15 A: If the gravitational field strength is 3.7 N/kg, and the space vehicle weighs 1200 kg, the gravitational force on the space vehicle must be 1200kg × (3.7 N/kg) = 4440N. If there's a downward force of 4440N due to gravity, the normal force must be equal and opposite, or 4440N upward, therefore the best answer is (3) 4400N.

4.16 Q: Which body is in equilibrium?

(1) a satellite orbiting Earth in a circular orbit

(2) a ball falling freely toward the surface of Earth

(3) a car moving with constant speed along a straight, level road

(4) a projectile at the highest point in its trajectory

4.16 A: (3) a car moving with constant speed.

Newton's 3rd Law

Newton's 3rd Law of Motion, commonly referred to as the Law of Action / Reaction, describes the phenomena by which all forces come in pairs. If Object 1 exerts a force on Object 2, then Object 2 must exert a force back on Object 1. Moreover, the force of Object 1 on Object 2 is equal in magnitude, or size, but opposite in direction to the force of Object 2 on Object 1. Written mathematically:

$$\vec{F}_{1on2} = -\vec{F}_{2on1}$$

This has many implications, many of which aren't immediately obvious. For example, if you punch the wall with your fist with a force of 100N, the wall imparts a force back on your fist of 100N (which is why it hurts!). Or try this. Push on the corner of your desk with your palm for a few seconds. Now look at your palm... see the indentation? That's because the corner of the desk pushed back on your palm.

Although this law surrounds our actions everyday, often times we don't even realize its effects. To run forward, a cat pushes with its legs backward on the ground, and the ground pushes the cat forward. How do you swim? If you want to swim forwards, which way do you push on the water? Backwards, that's right. As you push backwards on the water, the reactionary force, the water pushing you, propels you forward. How do you jump up in the air? You push down on the ground, and it's the reactionary force of the ground pushing on you that accelerates you skyward!

As you can see, then, forces always come in pairs. These pairs are known as **action-reaction pairs**. What are the action-reaction force pairs for a girl kicking a soccer ball? The girl's foot applies a force on the ball, and the ball applies an equal and opposite force on the girl's foot.

How does a rocket ship maneuver in space? The rocket propels hot expanding gas particles outward, so the gas particles in return push the rocket forward. Newton's 3rd Law even applies to gravity. The Earth exerts a gravitational force on you (downward). You, therefore, must apply a gravitational force upward on the Earth!

4.17 Q: Earth's mass is approximately 81 times the mass of the Moon. If Earth exerts a gravitational force of magnitude F on the Moon, the magnitude of the gravitational force of the Moon on Earth is

(1) F

(2) F/81

(3) 9F

(4) 81F

4.17 A: (1) The force Earth exerts on the Moon is the same in magnitude and opposite in direction of the force the Moon exerts on Earth.

4.18 Q: A 400-newton girl standing on a dock exerts a force of 100 newtons on a 10,000-newton sailboat as she pushes it away from the dock. How much force does the sailboat exert on the girl?

(1) 25 N

(2) 100 N

(3) 400 N

(4) 10,000 N

4.18 A: (2) The force the girl exerts on the sailboat is the same in magnitude and opposite in direction of the force the sailboat exerts on the girl.

4.19 Q: A carpenter hits a nail with a hammer. Compared to the magnitude of the force the hammer exerts on the nail, the magnitude of the force the nail exerts on the hammer during contact is

(1) less

(2) greater

(3) the same

4.19 A: (3) the same per Newton's 3rd Law.

Friction

Up until this point, we've been ignoring one of the most useful and most troublesome forces we deal with every day... a force that has tremendous application in transportation, machinery, and all parts of mechanics, yet we spend tremendous amounts of money each day fighting it. This force, **friction**, is a force that opposes motion.

4.20 Q: A projectile launched at an angle of 45° above the horizontal travels through the air. Compared to the projectile's theoretical path with no air friction, the actual trajectory of the projectile with air friction is

(1) lower and shorter

(2) lower and longer

(3) higher and shorter

(4) higher and longer

4.20 A: (1) lower and shorter. Friction opposes motion.

4.21 Q: A box is pushed toward the right across a classroom floor. The force of friction on the box is directed toward the

(1) left

(2) right

(3) ceiling

(4) floor

4.21 A: (1) left. Friction opposes motion.

There are two main types of friction. **Kinetic friction** is a frictional force that opposes motion for an object which is sliding along another surface. **Static friction**, on the other hand, acts on an object that isn't sliding. If you push on your textbook, but not so hard that it slides along your desk, static friction is opposing your applied force on the book, leaving the book in static equilibrium.

The magnitude of the frictional force depends upon two factors:

1. The nature of the surfaces in contact.
2. The normal force acting on the object (F_N).

The ratio of the frictional force and the normal force provides us with the **coefficient of friction** (μ), a proportionality constant that is specific to the two materials in contact.

You can look up the coefficient of friction for various surfaces on the front page of your Regents Physics Reference Table. Make sure you choose the appropriate coefficient. Use the static coefficient ($μ_s$) for objects which are not sliding, and the kinetic coefficient ($μ_k$) for objects which are sliding.

Approximate Coefficients of Friction		
	Kinetic	Static
Rubber on concrete (dry)	0.68	0.90
Rubber on concrete (wet)	0.58	
Rubber on asphalt (dry)	0.67	0.85
Rubber on asphalt (wet)	0.53	
Rubber on ice	0.15	
Waxed ski on snow	0.05	0.14
Wood on wood	0.30	0.42
Steel on steel	0.57	0.74
Copper on steel	0.36	0.53
Teflon on Teflon	0.04	

Which coefficient would you use for a sled sliding down a snowy hill? The kinetic coefficient of friction, of course. How about a refrigerator on your linoleum floor that is at rest and you want to start in motion? That would be the static coefficient of friction. Let's try a harder one... A car drives with its tires rolling freely. Is the friction between the tires and the road static or kinetic? Static. The tires are in constant contact with the road, much like walking. If the car was skidding, however, and the tires were locked, we would look at kinetic friction. Let's take a look at a sample problem:

4.22 Q: A car's performance is tested on various horizontal road surfaces. The brakes are applied, causing the rubber tires of the car to slide along the road without rolling. The tires encounter the greatest force of friction to stop the car on

(1) dry concrete

(2) dry asphalt

(3) wet concrete

(4) wet asphalt

4.22 A: To obtain the greatest force of friction (F_f), we'll need the greatest coefficient of friction (μ). Use the kinetic coefficient of friction (μ_k) since the tires are sliding. From the Approximate Coefficients of Friction table, the highest kinetic coefficient of friction for rubber comes from rubber on dry concrete. Answer: (1).

4.23 Q: The diagram below shows a block sliding down a plane inclined at angle θ with the horizontal.

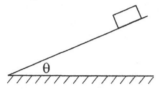

As angle θ is increased, the coefficient of kinetic friction between the bottom surface of the block and the surface of the incline will

(1) decrease

(2) increase

(3) remain the same

4.23 A: (3) remain the same. Coefficient of friction depends only upon the materials in contact.

The normal force always acts perpendicular to a surface, and comes from the interaction between atoms that act to maintain its shape. In many cases, it can be thought of as the elastic force trying to keep a flat surface flat (instead of bowed). We'll use the normal force to help us calculate the magnitude of the frictional force.

The force of friction, depending only upon the nature of the surfaces in contact (μ) and the magnitude of the normal force (F_N), therefore, can be determined using the formula:

$$F_f = \mu F_N$$

Solving problems involving friction requires us to apply the same basic principles we've been talking about throughout the dynamics unit... drawing a free body diagram, applying Newton's 2nd Law along the x- and/or y-axes, and solving for our unknowns. The only new skill is drawing the frictional force on the free body diagram, and using the relationship between the force of friction and the normal force to help us solve for our unknowns. Let's take a look at another sample problem:

4.24 Q: The diagram below shows a 4.0-kilogram object accelerating at 10 meters per second² on a rough horizontal surface.

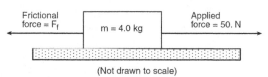

Acceleration = 10. m/s² ⟶

Frictional force = F_f   m = 4.0 kg   Applied force = 50. N

(Not drawn to scale)

What is the magnitude of the frictional force F_f acting on the object?

(1) 5.0 N

(2) 10 N

(3) 20 N

(4) 40 N

4.24 A: Define to the right as the positive direction.

$$F_{net} = ma$$

$$F_{app} - F_f = ma$$

$$F_f = F_{app} - ma$$

$$F_f = 50N - (4kg)(10\,{}^m\!/_{s^2}) = 10N$$

Answer: (2) 10 N.

Let's take a look at a more involved problem, tying together free body diagrams, Newton's 2nd Law, and the coefficient of friction:

4.25 Q: An ice skater applies a horizontal force to a 20-kilogram block on frictionless, level ice, causing the block to accelerate uniformly at 1.4 m/s² to the right. After the skater stops pushing the block, it slides onto a region of ice that is covered by a thin layer of sand. The coefficient of kinetic friction between the block and the sand-covered ice is 0.28. Calculate the magnitude of the force applied to the block by the skater.

4.25 A: Define right as the positive direction.

$$F_{net} = ma$$

$$F_{net} = (20kg)(1.4\,{}^m\!/_s) = 28N$$

4.26 Q: Referring to the problem of 4.25, determine the magnitude of the normal force acting on the block.

4.26 A: $F_{net_y} = ma_y$

$F_{net_y} = F_N - mg = 0$

$F_N = mg = (20kg)(9.81\,{}^m\!/_{s^2}) = 196N$

4.27 Q: Referring to the problem of 4.25, calculate the magnitude of the force of friction acting on the block as it slides over the sand-covered ice.

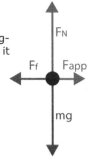

4.27 A: $F_f = \mu F_N$

$F_f = (0.28)(196N) = 55N$

These same steps can be used in many different ways in many different problems, but the same basic problem solving methodology still works... draw a free body diagram, apply Newton's 2nd Law, utilize the friction formula if necessary, and solve!

4.28 Q: A horizontal force of 8.0 newtons is used to pull a 20-newton wooden box moving toward the right along a horizontal, wood surface, as shown.

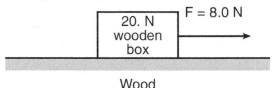

Calculate the magnitude of the frictional force acting on the box.

4.28 A: Recognize that the box has a weight of 20N, therefore its normal force must be 20N since it is not accelerating vertically.

$F_f = \mu F_N = (0.30)(20N) = 6N$

4.29 Q: Referring to the problem of 4.28, determine the magnitude of the net force acting on the box.

4.29 A: $F_{net} = F_{app} - F_f = 8N - 6N = 2N$

4.30 Q: Referring to the problem of 4.28, determine the mass of the box.

4.30 A:　$mg = 20N$

$$m = \frac{20N}{g} = \frac{20N}{9.81 \, ^m/_{s^2}} = 2.04kg$$

4.31 Q:　Referring to the problem of 4.28, calculate the magnitude of the acceleration of the box.

4.31 A:　$a = \dfrac{F_{net}}{m} = \dfrac{2N}{2.04kg} = 0.98 \, ^m/_{s^2}$

4.32 Q:　Compared to the force needed to start sliding a crate across a rough level floor, the force needed to keep it sliding once it is moving is

(1) less

(2) greater

(3) the same

4.32 A:　(1) less, since kinetic friction is less than static friction.

4.33 Q:　An airplane is moving with a constant velocity in level flight. Compare the magnitude of the forward force provided by the en- gines to the magnitude of the backward frictional drag force.

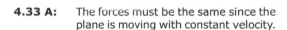

4.33 A:　The forces must be the same since the plane is moving with constant velocity.

4.34 Q:　When a 12-newton horizontal force is applied to a box on a hori- zontal tabletop, the box remains at rest. The force of static fric- tion acting on the box is

(1) 0 N

(2) between 0 N and 12 N

(3) 12 N

(4) greater than 12 N

4.34 A:　(3) 12 N. Because the box is at rest in static equilibrium, all forces on it must be balanced, therefore the force of static friction must be 12N.

Ramps and Inclines

Now that we've developed an understanding of Newton's Laws of Motion, free body diagrams, friction, and forces on flat surfaces, we can extend these tools to situations on ramps, or inclined surfaces. The key to understanding these situations is creating an accurate free body diagram after choosing convenient x- and y-axes. Problem-solving steps are consistent with those developed for Newton's 2nd Law.

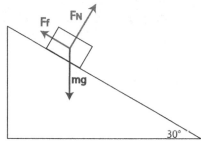

Let's take the example of a box on a ramp inclined at an angle of θ with respect to the horizontal. We can draw a basic free body diagram for this situation, with the force of gravity pulling the box straight down, the normal force perpendicular out of the ramp, and friction opposing motion (in this case pointing up the ramp).

Once the forces acting on the box have been identified, we must be clever about our choice of x-axis and y-axis directions. Much like we did when analyzing free falling objects and projectiles, if we set the positive x-axis in the direction of initial motion (or the direction the object wants to move if it is not currently moving), the y-axis must lie perpendicular to the ramp's surface (parallel to the normal force). Let's re-draw our free body diagram, this time superimposing it on our new axes.

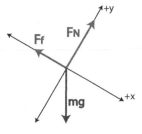

Unfortunately, the force of gravity on the box, *mg*, doesn't lie along one of the axes. Therefore, it must be broken up into components which do lie along the x- and y-axes in order to simplify our mathematical analysis. To do this, we can use geometry to break the weight down into a component parallel with the axis of motion (mg∥) and a component perpendicular to the x-axis (mg⊥) using the equations:

$$mg_{\parallel} = mg\sin(\theta)$$

$$mg_{\perp} = mg\cos(\theta)$$

Using these equations, we can re-draw the free body diagram, replacing *mg* with its components. Now all the forces line up with the axes, making it straightforward to write Newton's 2nd Law Equations (F_{NETx} and F_{NETy}) and continue with our standard problem-solving strategy.

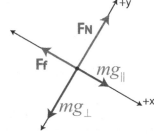

In the example shown with our modified free body diagram, we could write our Newton's 2nd Law Equations for both the x- and y-directions as follows:

$$F_{net_x} = mg_\parallel - F_f = mg\sin(\theta) - F_f = ma_x$$
$$F_{net_y} = F_N - mg_\perp = F_N - mg\cos(\theta) = 0$$

From this point, our problem becomes an exercise in algebra. If you need to tie the two equations together to eliminate a variable, don't forget the equation for the force of friction.

4.35 Q: Three forces act on a box on an inclined plane as shown in the diagram below. [Vectors are not drawn to scale.]

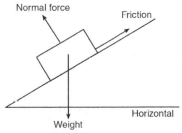

If the box is at rest, the net force acting on it is equal to

(1) the weight
(2) the normal force
(3) friction
(4) zero

4.35 A: (4) zero. If the box is at rest, the acceleration must be zero, therefore the net force must be zero.

4.36 Q: A 5-kg mass is held at rest on a frictionless 30° incline by force F. What is the magnitude of F?

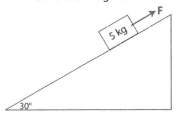

4.36 A: Start by identifying the forces on the box and making a free body diagram.

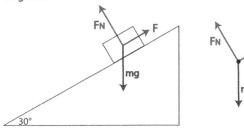

Break up the weight of the box into components parallel to and perpendicular to the ramp, and re-draw the free body diagram using the components of the box's weight.:

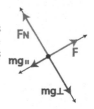

$$mg_{\parallel} = mg\sin(\theta)$$

$$mg_{\perp} = mg\cos(\theta)$$

Finally, use Newton's 2nd Law in the x-direction to solve for the force F.

$$F_{net} = F - mg_{\parallel} = F - mg\sin(\theta) = 0$$

$$F = mg\sin(\theta)$$

$$F = (5kg)(9.81\,\sfrac{m}{s^2})\sin(30°) = 24.5N$$

4.37 Q: A 10-kg box slides down a frictionless 18° ramp. Find the acceleration of the box, and the time it takes the box to slide 2 meters down the ramp.

4.37 A: Start by identifying the forces on the box and making a free body diagram.

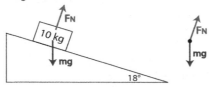

Break up the weight of the box into components parallel to and perpendicular to the ramp.

$$mg_{\parallel} = mg\sin(\theta)$$

$$mg_{\perp} = mg\cos(\theta)$$

Use Newton's 2nd Law to find the acceleration.

$$F_{net} = mg_{\parallel} = mg\sin(\theta) = ma$$

$$a = g\sin(\theta) = (9.81\,\sfrac{m}{s^2})(\sin(18°)) = 3.03\,\sfrac{m}{s^2}$$

Finally, use the acceleration to solve for the time it taxes the box to travel 2m down the ramp using kinematic equations.

$$d = v_i t + \tfrac{1}{2}at^2 = (0) + \tfrac{1}{2}at^2$$

$$t = \sqrt{\frac{2d}{a}} = \sqrt{\frac{2(2m)}{3.03\,\sfrac{m}{s^2}}} = 1.15s$$

4.38 Q: The diagram below shows a 1.0 × 10⁵-newton truck at rest on a hill that makes an angle of 8.0° with the horizontal.

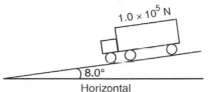

What is the component of the truck's weight parallel to the hill?
(1) 1.4 × 10³-newton
(2) 1.0 × 10⁴-newton
(3) 1.4 × 10⁴-newton
(4) 9.9 × 10⁴-newton

4.38 A: (3) 1.4 × 10⁴-newton

$$mg_{\parallel} = mg\sin(\theta) = (1.0 \times 10^5\,N)(\sin(8^\circ)) = 1.4 \times 10^4\,N$$

4.39 Q: A block weighing 10 newtons is on a ramp inclined at 30° to the horizontal. A 3-newton force of friction, F_f , acts on the block as it is pulled up the ramp at constant velocity with force F, which is parallel to the ramp, as shown in the diagram below.

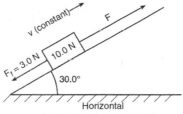

(Vectors not drawn to scale)

What is the magnitude of force F?
(1) 7 N
(2) 8 N
(3) 10 N
(4) 13 N

4.39 A: (2) Draw FBD and break weight of the box up into components.

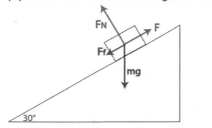

Break up the weight of the box into components parallel to and perpendicular to the ramp, then use Newton's second law, realizing that the forces must be balanced (net force is zero) since the box is moving at constant velocity.

$$F_{net} = F - F_f - mg\sin(\theta) = 0$$
$$F = F_f + mg\sin(\theta) = 3N + 10N \times \sin(30°) = 8N$$

4.40 Q: Which vector diagram best represents a cart slowing down as it travels to the right on a horizontal surface?

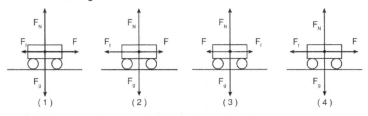

(1)   (2)   (3)   (4)

4.40 A: (2) Vertical forces are balanced, net force horizontally is in opposite direction of motion to create a negative acceleration and slow the cart down.

4.41 Q: The diagram represents a block at rest on an incline. Which diagram best represents the forces acting on the block? (F_f = frictional force, F_N = normal force, and F_w = weight.)

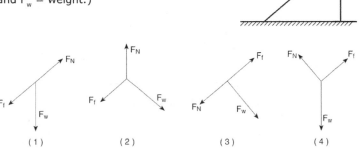

(1)   (2)   (3)   (4)

4.41 A: Correct Answer is (4).

You can find more practice problems on the APlusPhysics website at: http://www.aplusphysics.com/regents.

Chapter 5:
Circular Motion & Gravity

"I can calculate the motion of heavenly bodies, but not the madness of people."

— Sir Isaac Newton

Objectives

1. Explain the acceleration of an object moving in a circle at constant speed.
2. Define centripetal force and recognize that it is not a special kind of force, but that it is provided by forces such as tension, gravity, and friction.
3. Solve problems involving calculations of centripetal force.
4. Determine the direction of a centripetal force and centripetal acceleration for an object moving in a circular path.
5. Calculate the period, frequency, speed and distance traveled for objects moving in circles at constant speed.
6. Analyze and solve problems involving objects moving in vertical circles.
7. Determine the acceleration due to gravity near the surface of Earth.
8. Utilize Newton's Law of Universal Gravitation to determine the gravitational force of attraction between two objects.
9. Explain the difference between mass and weight.
10. Explain weightlessness for objects in orbit.

Now that we've talked about linear and projectile kinematics, as well as fundamentals of dynamics and Newton's Laws, we have the skills and background to analyze circular motion. Of course, this has the obvious applications such as cars moving around a circular track, roller coasters moving in loops, and toy trains chugging around a circular track under the Christmas tree. Less obvious, however, is the application to turning objects. Any object that is turning can be thought of as moving through a portion of a circular path, even if it's just a small portion of that path.

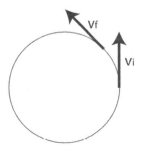

With this realization, analysis of circular motion will allow us to explore a car speeding around a corner on an icy road, a running back cutting upfield to avoid a blitzing linebacker, and the orbits of planetary bodies. The key to understanding all of these phenomena starts with the study of uniform circular motion.

Centripetal Acceleration

The motion of an object in a circular path at constant speed is known as **uniform circular motion** (UCM). An object in UCM is constantly changing direction, and since velocity is a vector and has direction, you could say that an object undergoing UCM has a constantly changing velocity, even if its speed remains constant. And if the velocity of an object is changing, it must be accelerating. Therefore, an object undergoing UCM is constantly accelerating. This type of acceleration is known as **centripetal acceleration**.

5.01 Q: If a car is accelerating, is its speed increasing?

5.01 A: It depends. Its speed could be increasing, or it could be accelerating in a direction opposite its velocity (slowing down). Or, its speed could remain constant yet still be accelerating if it is traveling in uniform circular motion.

Just as importantly, we need to figure out the direction of the object's acceleration, since acceleration is a vector. To do this, let's draw an object moving counter-clockwise in a circular path, and show its velocity vector at two different points in time. Since we know acceleration is the rate of change of an object's velocity with respect to time, we can determine the direction of the object's acceleration by finding the direction of its change in velocity, Δv.

Chapter 5: Circular Motion & Gravity

To find its change in velocity, Δv, we must recall that $\Delta v = v_f - v_i$.

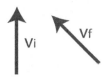

Therefore, we need to find the difference of the vectors v_f and v_i graphically, which can be re-written as $\Delta v = v_f + (-v_i)$.

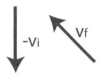

Recall that to add vectors graphically, we line them up tip-to-tail, then draw the resultant vector from the starting point (tail) of the first vector to the ending point (tip) of the last vector.

So, the acceleration vector must point in the direction shown above. If this vector is shown back on the original circle, lined up directly between the initial and final velocity vector, it's easy to see that the acceleration vector points toward the center of the circle.

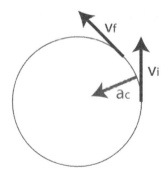

You can repeat this procedure from any point on the circle... no matter where you go, the acceleration vector always points toward the center of the circle. In fact, the word centripetal in centripetal acceleration means "center-seeking!"

So now we know the direction of an object's acceleration (toward the center of the circle), but what about its magnitude? The formula for the magnitude of an object's centripetal acceleration (which can be found on the Regents Physics Reference Table) is given by:

$$a_c = \frac{v^2}{r}$$

5.02 Q: In the diagram below, a cart travels clockwise at constant speed in a horizontal circle.

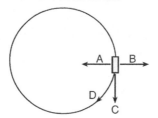

At the position shown in the diagram, which arrow indicates the direction of the centripetal acceleration of the cart?

(1) A

(2) B

(3) C

(4) D

5.02 A: (1) The acceleration of any object moving in a circlular path is toward the center of the circle.

5.03 Q: The diagram shows the top view of a 65-kilogram student at point A on an amusement park ride. The ride spins the student in a horizontal circle of radius 2.5 meters, at a constant speed of 8.6 meters per second. The floor is lowered and the student remains against the wall without falling to the floor.

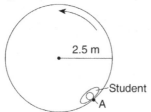

Which vector best represents the direction of the centripetal acceleration of the student at point A?

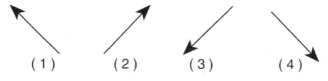

5.03 A: (1) Centripetal acceleration points toward the center of the circle.

5.04 Q: Which graph best represents the relationship between the magnitude of the centripetal acceleration and the speed of an object moving in a circle of constant radius?

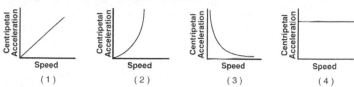

5.04 A: (2) Centripetal acceleration is proportional to v^2/r.

5.05 Q: A car rounds a horizontal curve of constant radius at a constant speed. Which diagram best represents the directions of both the car's velocity, v, and acceleration, a?

5.05 A: (3) Velocity is tangent to the circular path, and acceleration is toward the center of the circular path.

5.06 Q: A 0.50-kilogram object moves in a horizontal circular path with a radius of 0.25 meter at a constant speed of 4.0 meters per second. What is the magnitude of the object's acceleration?

(1) 8 m/s²

(2) 16 m/s²

(3) 32 m/s²

(4) 64 m/s²

5.06 A: (4) 64 m/s².

Circular Speed

So how do you find the speed of an object as it travels in a circular path? The formula for speed that we learned in kinematics still applies.

$$\overline{v} = \frac{d}{t}$$

We have to be careful in using this equation, however, to understand that an object traveling in a circular path is traveling along the circumference of a circle. Therefore, if an object were to make one complete revolution around the circle, the distance it travels is equal to the circle's circumference.

$$C = 2\pi r$$

5.07 Q: Miranda drives her car clockwise around a circular track of radius 30m. She completes 10 laps around the track in 2 minutes. Find Miranda's total distance traveled, average speed, and centripetal acceleration.

5.07 A: $d = 10 \times 2\pi r = 10 \times 2\pi(30m) = 1885m$

$$\bar{v} = \frac{d}{t} = \frac{1885m}{120s} = 15.7 \, ^m/_s$$

$$a_c = \frac{v^2}{r} = \frac{(15.7 \, ^m/_s)^2}{30m} = 8.2 \, ^m/_{s^2}$$

5.08 Q: The combined mass of a race car and its driver is 600 kilograms. Traveling at constant speed, the car completes one lap around a circular track of radius 160 meters in 36 seconds. Calculate the speed of the car.

5.08 A: $\bar{v} = \frac{d}{t} = \frac{2\pi r}{t} = \frac{2\pi(160m)}{36s} = 27.9 \, ^m/_s$

Centripetal Force

If an object traveling in a circular path has an inward acceleration, Newton's 2nd Law tells us there must be a net force directed toward the center of the circle as well. This type of force, known as a **centripetal force**, can be a gravitational force, a tension, an applied force, or even a frictional force.

NOTE: When dealing with circular motion problems, it is important to realize that a centripetal force isn't really a new force, a centripetal force is just a label or grouping we apply to a force to indicate its direction is toward the center of a circle. This means that you never want to label a force on a free body diagram as a centripetal force, F_c. Instead, label the center-directed force as specifically as you can. If a tension is causing the force, label the force F_T. If a frictional force is causing the center-directed force, label it F_f, and so forth.

We can combine the equation for centripetal acceleration with Newton's 2nd Law to obtain Newton's 2nd Law for Circular Motion. Recall that Newton's 2nd Law states:

$$F_{net} = ma$$

For an object traveling in a circular path, there must be a net (centripetal) force directed toward the center of the circular path to cause a (centripetal) acceleration directed toward the center of the circular path. We can revise Newton's 2nd Law for this particular case, then, as follows:

$$F_C = ma_C$$

Then, recalling the formula for centripetal acceleration as:

$$a_c = \frac{v^2}{r}$$

We can put these together, replacing a_c in our equation to get a combined form of Newton's 2nd Law for Uniform Circular Motion:

$$F_C = \frac{mv^2}{r}$$

Of course, if an object is traveling in a circular path and the centripetal force is removed, the object will continue traveling in a straight line in whatever direction it was moving at the instant the force was removed.

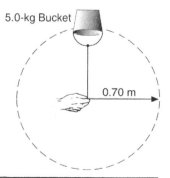

5.09 Q: An 800N running back turns a corner in a circular path of radius 1 meter at a velocity of 8 m/s. Find the running back's mass, centripetal acceleration, and centripetal force.

5.09 A: $mg = 800N \Rightarrow m = \dfrac{800N}{9.81^{m}/_{s^2}} = 81.5kg$

$a_C = \dfrac{v^2}{r} = \dfrac{(8^{m}/_{s})^2}{1m} = 64^{m}/_{s^2}$

$F_C = ma_C = (81.5kg)(64^{m}/_{s^2}) = 5220N$

5.10 Q: The diagram at right shows a 5.0-kilogram bucket of water being swung in a horizontal circle of 0.70-meter radius at a constant speed of 2.0 meters per second. The magnitude of the centripetal force on the bucket of water is approximately

(1) 5.7 N

(2) 14 N

(3) 29 N

(4) 200 N

5.10 A: (3) $F_c = ma_c = m\dfrac{v^2}{r} = (5kg)\dfrac{(2\,{}^m\!/\!_s)^2}{0.7m} = 29N$

5.11 Q: A 1.0×10^3-kilogram car travels at a constant speed of 20 meters per second around a horizontal circular track. Which diagram correctly represents the direction of the car's velocity (v) and the direction of the centripetal force (F_c) acting on the car at one particular moment?

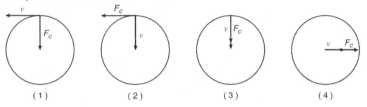

      (1)   (2)   (3)   (4)

5.11 A: (1) Velocity is tangent to the circle, and the centripetal force points toward the center of the circle.

5.12 Q: A 1750-kilogram car travels at a constant speed of 15 meters per second around a horizontal, circular track with a radius of 45 meters. The magnitude of the centripetal force acting on the car is

(1) 5.00 N

(2) 583 N

(3) 8750 N

(4) 3.94 10^5 N

5.12 A: (3) $F_c = ma_c = m\dfrac{v^2}{r} = (1750kg)\dfrac{(15\,{}^m\!/\!_s)^2}{45m} = 8750N$

5.13 Q: A ball attached to a string is moved at constant speed in a horizontal circular path. A target is located near the path of the ball as shown in the diagram. At which point along the ball's path should the string be released, if the ball is to hit the target?

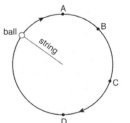

(1) A

(2) B

(3) C

(4) D

5.13 A: (B) If released at point B, the ball will continue in a straight line to the target.

5.14 Q: A 1200-kilogram car traveling at a constant speed of 9 meters per second turns at an intersection. The car follows a horizontal circular path with a radius of 25 meters to point P. At point P, the car hits an area of ice and loses all frictional force on its tires. Which path does the car follow on the ice?

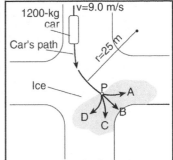

(1) A

(2) B

(3) C

(4) D

5.14 A: (2) Once the car loses all frictional force, there is no longer a force toward the center of the circular path, therefore the car will travel in a straight line toward B.

Frequency and Period

For objects moving in circular paths, we can characterize their motion around the circle using the terms frequency (f) and period (T). The **frequency** of an object is the number of revolutions the object makes in a complete second. It is measured in units of [1/s], or Hertz (Hz). In similar fashion, the **period** of an object is the time it takes to make one complete revolution. Since the period is a time interval, it is measured in units of seconds. We can relate period and frequency using the equations:

$$f = \frac{1}{T} \qquad T = \frac{1}{f}$$

5.15 Q: A 500g toy train completes 10 laps of its circular track in 1 minute and 40 seconds. If the diameter of the track is 1 meter, find the train's centripetal acceleration (a_c), centripetal force (F_c), period (T), and frequency (f).

5.15 A: $\overline{v} = \dfrac{d}{t} = \dfrac{2\pi r \times 10}{t} = \dfrac{2\pi(0.5m) \times 10}{100s} = 0.314\,{}^m\!/_s$

$a_c = \dfrac{v^2}{r} = \dfrac{(0.314\,{}^m\!/_s)^2}{0.5m} = 0.197\,{}^m\!/_{s^2}$

$F_c = ma_c = (0.5kg)(0.197\,{}^m\!/_{s^2}) = 0.099\,N$

$T = \dfrac{100s}{10revs} = 10s$

$f = \dfrac{1}{T} = \dfrac{1}{10s} = 0.1Hz$

5.16 Q: Alan makes 38 complete revolutions on the playground Round-A-Bout in 30 seconds. If the radius of the Round-A-Bout is 1 meter, determine

(A) Period of the motion

(B) Frequency of the motion

(C) Speed at which Alan revolves

(D) Centripetal force on 40-kg Alan

5.16 A: (A) $T = \dfrac{30s}{38revs} = 0.789s$

(B) $f = \dfrac{1}{T} = \dfrac{1}{0.789s} = 1.27\,Hz$

(C) $\overline{v} = \dfrac{d}{t} = \dfrac{38 \times 2\pi r}{t} = \dfrac{38 \times 2\pi(1m)}{30s} = 7.96\,{}^m\!/_s$

(D) $F_c = ma_c = m\dfrac{v^2}{r} = (40kg)\dfrac{(7.96\,{}^m\!/_s)^2}{1m} = 2530\,N$

Vertical Circular Motion

Objects travel in circles vertically as well as horizontally. Because the speed of these objects isn't typically constant, technically this isn't uniform circular motion, but our UCM analysis skills still prove applicable.

Consider a roller coaster traveling in a vertical loop of radius 10m. You travel through the loop upside down, yet you don't fall out of the roller coaster. How is this possible? We can use our understanding of UCM and dynamics to find out!

To begin with, let's first take a look at the coaster when the car is at the bottom of the loop. Drawing a free body diagram, the force of gravity on the coaster, also known as its weight, pulls it down, so we draw a vector pointing down labeled "mg." Opposing that force is the normal force of the rails of the coaster pushing up, which we label F_N.

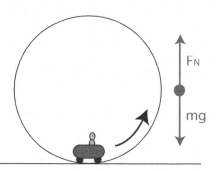

Because the coaster is moving in a circular path, we can analyze it using the tools we developed for uniform circular motion. Newton's 2nd Law still applies, so we can write:

$$F_{NET_C} = F_N - mg$$

Notice that because we're talking about circular motion, we'll adopt the convention that forces pointing toward the center of the circle are positive, and forces pointing away from the center of the circle are negative. At this point, recall that the force you "feel" when you're in motion is actually the normal force. So, solving for the normal force as you begin to move in a circle, we find:

$$F_N = F_{NET_C} + mg$$

Since we know that the net force is always equal to mass times acceleration, the net centripetal force is equal to mass times the centripetal acceleration, so we can replace F_{NETc} as follows:

$$F_N = F_{NET_C} + mg = \frac{mv^2}{r} + mg$$

We can see from the resulting equation that the normal force is now equal to the weight plus an additional term from the centripetal force of the circular motion. As we travel in a circular path near the bottom of the loop, we feel heavier than our weight. In common terms, we feel additional "g-forces." How many g's we feel can be obtained with a little bit more manipulation. If we re-write our equation for the normal force, pulling out the mass by applying the distributive property of multiplication, we obtain:

$$F_N = m\left(\frac{v^2}{r} + g\right)$$

Notice that inside the parenthesis we have our standard acceleration due to gravity, g, plus a term from the centripetal acceleration. This additional term is the additional g-force felt by a person. For example, if a_c was equal to g (9.81 m/s²), you could say the person in the cart was experiencing two g's (1g from the centripetal acceleration, and 1g from the Earth's gravitational field). If a_c were equal to 3×g (29.4 m/s²), the person would be experiencing a total of four g's.

Expanding this analysis to a similar situation in a different context, try to imagine instead of a roller coaster, a mass whirling in a vertical circle by a string. You could replace the normal force by the tension in the string in our analysis. Because the force is larger at the bottom of the circle, the likelihood of the string breaking is highest when the mass is at the bottom of the circle!

At the top of the loop, we have a considerably different picture. Now, the normal force from the coaster rails must be pushing down against the cart, though still in the positive direction since down is now toward the center of the circular path. In this case, however, the weight of the object also points toward the center of the circle, since the Earth's gravitational field always pulls toward the center of the Earth. Our free body diagram looks considerably different, and therefore our application to Newton's 2nd Law for Circular Motion is considerably different as well:

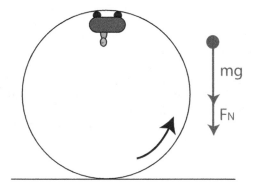

$$F_{NET_c} = F_N + mg$$

Since the force you feel is actually the normal force, we can solve for the normal force and expand the net centripetal force as shown:

$$F_N = F_{NET_c} - mg = \frac{mv^2}{r} - mg$$

Chapter 5: Circular Motion & Gravity

You can see from the equation that the normal force is now the centripetal force minus your weight. If the centripetal force were equal to your weight, you would feel as though you were weightless. Note that this is also the point where the normal force is exactly equal to 0. This means the rails of the track are no longer pushing on the roller coaster cart. If the centripetal force was slightly smaller, and the car's speed was slightly smaller, the normal force F_N would be less than 0. Since the rails can't physically pull the cart in the negative direction (away from the center of the circle), this means the car is falling off the rail and the cart's occupant is about to have a very, very bad day. Only by maintaining a high speed can the cart successfully negotiate the loop. Go too slow and the cart falls.

In order to remain safe, real roller coasters actually have wheels on both sides of the rails to prevent the cart from falling if it ever did slow down at the top of a loop, although coasters are designed so that this situation never actually occurs.

5.17 Q: In an experiment, a rubber stopper is attached to one end of a string that is passed through a plastic tube before weights are-attached to the other end. The stopper is whirled in a horizontal circular path at constant speed.

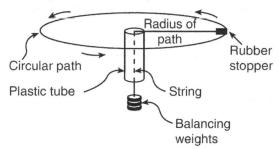

(A) Describe what would happen to the radius of the circle if the student whirls the stopper at a greater speed without changing the balancing weights.

(B) The rubber stopper is now whirled in a vertical circle at the same speed. On the vertical diagram, draw and label vectors to indicate the direction of the weight (F_g) and the direction of the centripetal force (F_c) at the position shown.

5.17 A: (A) As the speed of the stopper is increased, the radius of the orbit will increase.

(B)

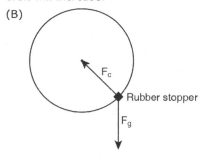

Universal Gravitation

All objects that have mass attract each other with a gravitational force. The magnitude of that force, F_g, can be calculated using Newton's Law of Universal Gravitation:

$$F_g = \frac{Gm_1 m_2}{r^2}$$

This law tells us that the force of gravity between two objects is proportional to each of the masses(m_1 and m_2) and inversely proportional to the square of the distance between them (r). The **universal gravitational constant**, G, is a "fudge factor," so to speak, included in the equation so that your answers come out in S.I. units. G is given on the front page of your Regents Physics Reference Table as 6.67×10^{-11} N·m²/kg².

Let's look at this relationship in a bit more detail. Force is directly proportional to the masses of the two objects, therefore if either of the masses were doubled, the gravitational force would also double. In similar fashion, if the distance between the two objects, r, was doubled, the force of gravity would be quartered since the distance is squared in the denominator. This type of relationship is called an inverse square law, which describes many phenomena in the natural world.

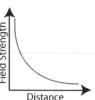

NOTE: The distance between the masses, r, is actually the distance between the center of masses of the objects. For large objects, such as the gravitational attraction between the Earth and the moon, for example, you must determine the distance from the center of the Earth to the center of the moon, not their surfaces.

Some hints for problem solving when dealing with Newton's Law of Universal Gravitation:

1. Substitute values in for variables at the very end of the problem only. The longer you can keep the formula in terms of variables, the fewer opportunities for mistakes.
2. Before using your calculator to find an answer, try to estimate the order of magnitude of the answer. Use this to check your final answer.
3. Once your calculations are complete, make sure your answer makes sense by comparing your answer to a known or similar quantity. If your answer doesn't make sense, check your work and verify your calculations.

5.18 Q: What is the gravitational force of attraction between two asteroids in space, each with a mass of 50,000 kg, separated by a distance of 3800 m?

5.18 A: $F_g = \dfrac{Gm_1 m_2}{r^2}$

$$F_g = \frac{(6.67 \times 10^{-11}\ ^{N\bullet m^2}\!/_{kg^2})(50000kg)(50000kg)}{(3800m)^2} = 1.15 \times 10^{-8}\ N$$

As you can see, the force of gravity is a relatively weak force, and we would expect a relatively weak force between relatively small objects. It takes tremendous masses and relatively small distances in order to develop significant gravitational forces. Let's take a look at another problem to explore the relationship between gravitational force, mass, and distance.

5.19 Q: As a meteor moves from a distance of 16 Earth radii to a distance of 2 Earth radii from the center of Earth, the magnitude of the gravitational force between the meteor and Earth becomes

(1) one-eighth as great

(2) 8 times as great

(3) 64 times as great

(4) 4 times as great

5.19 A: (3) 64 times as great. The gravitational force is given by Newton's Law of Universal Gravitation. If the radius is one-eighth its initial value, and radius is squared in the denominator, the radius squared becomes one-sixty fourth its initial value. Because radius squared is in the denominator, the gravitational force must increase by 64X.

5.20 Q: Which diagram best represents the gravitational forces, F_g, between a satellite, S, and Earth?

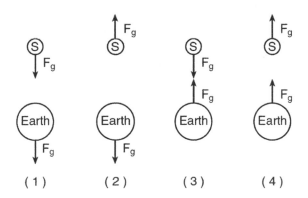

5.20 A: (3) Newton's 3rd Law says that the force of gravity on the satellite due to Earth will be equal in magnitude and opposite in the direction the force of gravity on the Earth due to the satellite.

5.21 Q: Io (pronounced "EYE oh") is one of Jupiter's moons discovered by Galileo. Io is slightly larger than Earth's Moon. The mass of Io is 8.93×10^{22} kilograms and the mass of Jupiter is 1.90×10^{27} kilograms. The distance between the centers of Io and Jupiter is 4.22×10^8 meters.

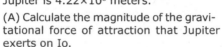

(A) Calculate the magnitude of the gravitational force of attraction that Jupiter exerts on Io.

(B) Calculate the magnitude of the acceleration of Io due to the gravitational force exerted by Jupiter.

5.21 A: (A) 6.35×10^{22} N

$$F_g = \frac{Gm_1 m_2}{r^2} = \frac{(6.67 \times 10^{-11} \, {}^{N \bullet m^2}\!/\!_{kg^2})(8.93 \times 10^{22} \, kg)(1.9 \times 10^{27} \, kg)}{(4.22 \times 10^8 \, m)^2}$$

$$F_g = 6.35 \times 10^{22} \, N$$

(B) $a = \dfrac{F_{net}}{m} = \dfrac{6.35 \times 10^{22} \, N}{8.93 \times 10^{22} \, kg} = 0.71 \, {}^{m}\!/\!_{s}$

5.22 Q: The diagram shows two bowling balls, A and B, each having a mass of 7 kilograms, placed 2 meters apart.

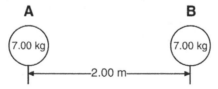

What is the magnitude of the gravitational force exerted by ball A on ball B?

(1) 8.17×10^{-9} N

(2) 1.63×10^{-9} N

(3) 8.17×10^{-10} N

(4) 1.17×10^{-10} N

5.22 A: (3) $F_g = \dfrac{Gm_1 m_2}{r^2} = \dfrac{(6.67 \times 10^{-11} \, {}^{N \bullet m^2}\!/\!_{kg^2})(7 kg)(7 kg)}{(2m)^2} = 8.17 \times 10^{-10} \, N$

5.23 Q: A 2.0-kilogram object is falling freely near Earth's surface. What is the magnitude of the gravitational force that Earth exerts on the object?

(1) 20 N

(2) 2.0 N

(3) 0.20 N

(4) 0.0 N

5.23 A: (1) 20 N

Gravitational Fields

Gravity is a non-contact, or field, force. Its effects are observed without the two objects coming into contact with each other. Exactly how this happens is a mystery to this day, but scientists have come up with a mental construct to help us understand how gravity works.

Envision an object with a gravitational field, such as the planet Earth. The closer other masses are to Earth, the more gravitational force they will experience. We can characterize this by calculating the amount of force the Earth will exert per unit mass at various distances from the Earth. Obviously, the closer the object is to the Earth, the larger a force it will experience, and the farther it is from the Earth, the smaller a force it will experience.

Attempting to visualize this, picture the strength of the gravitational force on a test object represented by a vector at the position of the object. The denser the force vectors are, the stronger the force, the stronger the "gravitational field." As these field lines become less dense, the gravitational field gets weaker.

To calculate the gravitational field strength at a given position, we can go back to our definition of the force of gravity on our test object, better known as its weight. We've been writing this as mg since we began our study of dynamics. Realizing that this is the force of gravity on an object, we can also calculate the force of gravity on a test mass using Newton's Law of Universal Gravitation. Putting these together we find that:

$$F_g = mg = \frac{Gm_1 m_2}{r^2}$$

Realizing that the mass on the left-hand side of the equation, the mass of our test object, is also one of the masses on the right-hand side of the equation, we can simplify our expression by dividing out the test mass.

$$g = \frac{Gm}{r^2}$$

Therefore, the gravitational field strength, g, is equal to the universal gravitational constant, G, times the mass of the object, divided by the square of the distance between the objects.

But wait, you might say, I thought g was the acceleration due to gravity on the surface of the Earth! And you would be right. Not only is g the gravitational field strength, it's also the acceleration due to gravity. The units even work out. The units of gravitational field strength, N/kg, are equivalent to the units for acceleration, m/s²!

Still skeptical? Let's calculate the gravitational field strength on the surface of the Earth using the knowledge that the mass of the Earth is approximately 5.98×10²⁴ kg and the distance from the surface to the center of mass of the Earth (which varies slightly since the Earth isn't a perfect sphere) is approximately 6378 km in New York.

$$g = \frac{Gm}{r^2} = \frac{(6.67 \times 10^{-11} \, {}^{N \bullet m^2}\!/_{kg^2})(5.98 \times 10^{24} \, kg)}{(6378000m)^2} = 9.81 \, {}^{m}\!/_{s^2}$$

As expected, the gravitational field strength on the surface of the Earth is the acceleration due to gravity.

5.24 Q: Suppose a 100-kg astronaut feels a gravitational force of 700N when placed in the gravitational field of a planet.

A) What is the gravitational field strength at the location of the astronaut?

B) What is the mass of the planet if the astronaut is 2×10⁶ m from its center?

5.24 A: (A) $F_g = mg$

$$g = \frac{F_g}{m} = \frac{700N}{100kg} = 7 \, {}^{N}\!/_{kg}$$

(B) $F_g = \frac{Gm_1 m_2}{r^2}$ $m_{planet} = \frac{F_g r^2}{Gm_{astronaut}}$

$$m_{planet} = \frac{(700N)(2 \times 10^6 \, m)^2}{(6.67 \times 10^{-11} \, {}^{N \bullet m^2}\!/_{kg^2})(100kg)} = 4.2 \times 10^{23} \, kg$$

5.25 Q: What is the acceleration due to gravity at a location where a 15-kilogram mass weighs 45 newtons?

(1) 675 m/s²

(2) 9.81 m/s²

(3) 3.00 m/s²

(4) 0.333 m/s²

5.25 A: (3) $F_g = mg$

$$g = \frac{F_g}{m} = \frac{45N}{15kg} = 3\,{}^N\!/_{kg} = 3\,{}^m\!/_{s^2}$$

5.26 Q: A 1200-kilogram space vehicle travels at 4.8 meters per second along the level surface of Mars. If the magnitude of the gravitational field strength on the surface of Mars is 3.7 newtons per kilogram, the magnitude of the normal force acting on the vehicle is

(1) 320 N

(2) 930 N

(3) 4400 N

(4) 5800 N

5.26 A: (3) To solve this problem, you must recognize that the gravitational field strength of 3.7 N/kg is equivalent to the acceleration due to gravity on Mars, therefore a=3.7 m/s². Then, because the space vehicle isn't accelerating vertically, the normal force must balance the vehicle's weight.

$$F_N = F_g = mg = (1200kg)(3.7\,{}^m\!/_s) = 4440N$$

5.27 Q: A 2.00-kilogram object weighs 19.6 newtons on Earth. If the acceleration due to gravity on Mars is 3.71 meters per second², what is the object's mass on Mars?

(1) 2.64 kg

(2) 2.00 kg

(3) 19.6 N

(4) 7.42 N

5.27 A: (2) 2.00 kg. Mass does not change.

5.28 Q: The graph below represents the relationship between gravitational force and mass for objects near the surface of Earth.

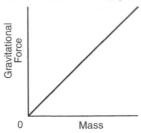

The slope of the graph represents the

(1) acceleration due to gravity

(2) universal gravitational constant

(3) momentum of objects

(4) weight of objects

5.28 A: (1) acceleration due to gravity.

Orbits

How do celestial bodies orbit each other? The moon orbits the Earth. The Earth orbits the sun. Our solar system is in orbit in the Milky Way galaxy... but how does it all work?

To explain orbits, Sir Isaac Newton developed a "thought experiment" in which he imagined a cannon placed on top of a very tall mountain, so tall, in fact, that the peak of the mountain was above the atmosphere (this is important because it allows us to neglect air resistance). If the cannon then launched a projectile horizontally, the projectile would follow a parabolic path to the surface of the Earth.

If the projectile was launched with a higher speed, however, it would travel farther across the surface of the Earth before reaching the ground. If its speed could be increased high enough, the projectile would fall at the same rate the Earth's surface curves away. The projectile would continue falling forever as it circled the Earth! This circular motion describes an orbit.

Put another way, the astronauts in the Space Shuttle aren't weightless. Far from it, actually, the Earth's gravity is still acting on them and pulling them toward the center of the Earth with a substantial force. We can even calculate that force.

5.29 Q: If the Space Shuttle orbits the Earth at an altitude of 380 km, what is the gravitational field strength due to the Earth?

5.29 A: Recall that we can obtain values for G, the mass of the Earth, and the radius of the Earth from the reference table.

$$F_g = mg = \frac{Gm_1m_2}{r^2} \rightarrow g = \frac{Gm_{Earth}}{r^2}$$

$$g = \frac{(6.67 \times 10^{-11}\ ^{N \bullet m^2}\!/_{kg^2})(5.98 \times 10^{24}\ kg)}{(6.37 \times 10^6\ m + 380 \times 10^3\ m)^2}$$

$$g = 8.75\ ^N\!/_{kg} = 8.75\ ^m\!/_{s^2}$$

This means that the acceleration due to gravity at the altitude the astronauts are orbiting the earth is only 11% less than on the surface of the Earth! In actuality, the Space Shuttle is falling, but it's moving so fast horizontally that by the time it falls, the Earth has curved away underneath it so that the shuttle remains at the same distance from the center of the Earth. It is in orbit! Of course, this takes tremendous speeds. To maintain an orbit of 380 km, the space shuttle travels approximately 7680 m/s, more than 23 times the speed of sound at sea level!

5.30 Q: Calculate the magnitude of the centripetal force acting on Earth as it orbits the Sun, assuming a circular orbit and an orbital speed of 3.00 × 10⁴ meters per second.

5.30 A: Use the reference table to find required constants.

$$F_c = ma_c = m\frac{v^2}{r}$$

$$F_c = (5.98 \times 10^{24}\ kg)\frac{(3 \times 10^4\ ^m\!/_s)^2}{(1.5 \times 10^{11}\ m)} = 3.59 \times 10^{22}\ N$$

5.31 Q: The diagram below represents two satellites of equal mass, A and B, in circular orbits around a planet.

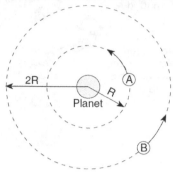

Compared to the magnitude of the gravitational force of attraction between satellite A and the planet, the magnitude of the gravitational force of attraction between satellite B and the planet is

(1) half as great

(2) twice as great

(3) one-fourth as great

(4) four times as great

5.31 A: (3) one-fourth as great due to the inverse square law relationship.

You can find more practice problems on the APlusPhysics website at: http://www.aplusphysics.com/regents.

Chapter 6: Momentum

"Sliding headfirst is the safest way to get to the next base, I think, and the fastest. You don't lose your momentum, and there's one more important reason I slide headfirst, it gets my picture in the paper."

— Pete Rose

Objectives

1. Define and calculate the momentum of an object.
2. Determine the impulse given to an object.
3. Use impulse to solve a variety of problems.
4. Interpret and use force vs. time graphs.
5. Apply conservation of momentum to solve a variety of problems.

We've talked in some depth now about motion, specifically trying to relate what we know about motion back to kinetic energy. Recall our definition of kinetic energy as the ability or capacity of a moving object to move another object. The key characteristics of kinetic energy, mass and velocity, can be observed from the equation:

$$KE = \tfrac{1}{2}mv^2$$

There's more to the story, however. We've talked about moving objects causing other objects to move, but we really haven't looked at those situations yet -- to do so we need to talk about collisions, and collisions are all about momentum.

Defining Momentum

Let's assume there's a car speeding toward you, out of control without its brakes, at a speed of 27 m/s (60 mph). Can you stop it by standing in front of it and holding out your hand? Why not?

Unless you're Superman, you probably don't want to try stopping a moving car by holding out your hand. It's too big, and it's moving way too fast. Attempting such a feat would result in a number of physics demonstrations upon your body, all of which would hurt.

We can't stop the car because it has too much momentum. **Momentum** is a vector quantity, given the symbol p, which measures how hard it is to stop a moving object. Of course, larger objects have more momentum than smaller objects, and faster objects have more momentum than slower objects. We can therefore calculate momentum using the equation:

$$p = mv$$

Momentum is the product of an object's mass times its velocity, and its units must be the same as the units of mass [kg] times velocity [m/s], therefore the units of momentum must be [kg·m/s], which can also be written as a Newton-second [N·s].

6.01 Q: Two trains, Big Red and Little Blue, have the same velocity. Big Red, however, has twice the mass of Little Blue. Compare their momenta.

6.01 A: Because Big Red has twice the mass of Little Blue, Big Red must have twice the momentum of Little Blue.

6.02 Q: The magnitude of the momentum of an object is 64 kilogram-meters per second. If the velocity of the object is doubled, the magnitude of the momentum of the object will be

(1) 32 kg·m/s

(2) 64 kg·m/s

(3) 128 kg·m/s

(4) 256 kg·m/s

6.02 A: (3) if velocity is doubled, momentum is doubled.

Because momentum is a vector, the direction of the momentum vector is the same as the direction of the object's velocity.

6.03 Q: An Aichi D3A bomber, with a mass of 3600 kg, departs from its aircraft carrier with a velocity of 85 m/s due east. What is the jet's momentum?

6.03 A: $p = mv = (3600kg)(85\,{}^{m}\!/_{s}) = 3.06 \times 10^{5}\,{}^{kg \cdot m}\!/_{s}$

Now, let's assume the jet drops its payload and has burned up most of its fuel as it continues its journey east to its destination air field.

6.04 Q: If the bomber's new mass is 3,000 kg, and due to its reduced weight the pilot increases the cruising speed to 120 m/s, what is the jet's new momentum?

6.04 A: $p = mv = (3000kg)(120\,{}^{m}\!/_{s}) = 3.60 \times 10^{5}\,{}^{kg \cdot m}\!/_{s}$

6.05 Q: Cart A has a mass of 2 kilograms and a speed of 3 meters per second. Cart B has a mass of 3 kilograms and a speed of 2 meters per second. Compared to the inertia and magnitude of momentum of cart A, cart B has

(1) the same inertia and a smaller magnitude of momentum

(2) the same inertia and the same magnitude of momentum

(3) greater inertia and a smaller magnitude of momentum

(4) greater inertia and the same magnitude of momentum

6.05 A: (4) greater inertia and the same magnitude of momentum.

Impulse

As you can see, momentum can change, and a change in momentum is known as an **impulse**. In Regents Physics, the vector quantity impulse is represented by a capital J, and since it's a change in momentum, its units are the same as those for momentum, [kg·m/s], and can also be written as a Newton-second [N·s].

$$J = \Delta p$$

6.06 Q: Assume the D3A bomber, which had a momentum of 3.6×10^5 kg·m/s, comes to a halt on the ground. What impulse is applied?

6.06 A: Define east as the positive direction:

$$J = \Delta p = p_f - p_i = 0 - 3.6 \times 10^5 \; {}^{kg \bullet m}\!/_{s^2}$$

$$J = -3.6 \times 10^5 \; {}^{kg \bullet m}\!/_{s^2} \; \text{east} = 3.6 \times 10^5 \; {}^{kg \bullet m}\!/_{s^2} \; \text{west}$$

6.07 Q: Calculate the magnitude of the impulse applied to a 0.75-kilogram cart to change its velocity from 0.50 meter per second east to 2.00 meters per second east.

6.07 A: $J = \Delta p = m\Delta v = (0.75 kg)(1.5 \, {}^{m}\!/_{s}) = 1.1 N \bullet s$

6.08 Q: A 6.0-kilogram block, sliding to the east across a horizontal, frictionless surface with a momentum of 30 kilogram•meters per second, strikes an obstacle. The obstacle exerts an impulse of 10 newton•seconds to the west on the block. The speed of the block after the collision is
(1) 1.7 m/s
(2) 3.3 m/s
(3) 5.0 m/s
(4) 20 m/s

6.08 A: (2) $J = \Delta p = m\Delta v = m(v_f - v_i) = mv_f - mv_i$

$$v_f = \frac{J = mv_i}{m} = \frac{(-10 N \bullet s) + 30 \, {}^{kg \bullet m}\!/_{s}}{6kg} = 3.3 \, {}^{m}\!/_{s}$$

6.09 Q: Which two quantities can be expressed using the same units?
(1) energy and force
(2) impulse and force
(3) momentum and energy
(4) impulse and momentum

6.09 A: (4) impulse and momentum both have units of kg·m/s.

6.10 Q: A 1000-kilogram car traveling due east at 15 meters per second is hit from behind and receives a forward impulse of 6000 newton-seconds. Determine the magnitude of the car's change in momentum due to this impulse.

6.10 A: Change in momentum is the definition of impulse, therefore the answer must be 6000 newton-seconds.

Impulse-Momentum Theorem

Since momentum is equal to mass times velocity, we can write that p=mv. We also know that impulse is a change in momentum, so impulse can be written as J=Δp. If we combine these equations, we find:

$$J = \Delta p = \Delta mv$$

Since the mass of a single object is constant, a change in the product of mass and velocity is equivalent to the product of mass and change in velocity. Specifically:

$$J = \Delta p = m\Delta v$$

A change in velocity is called acceleration. But what causes an acceleration? A force! And does it matter if the force is applied for a very short time or a very long time? Common sense says it does and also tells us that the longer the force is applied, the longer the object will accelerate, and therefore the greater the object's change in momentum!

We can prove this by using an old mathematician's trick -- if we multiply the right side of our equation by 1, we of course get the same thing. And if we multiply the right side of our equation by Δt/Δt, which is 1, we still get the same thing. Take a look:

$$J = \Delta p = \frac{m\Delta v\Delta t}{\Delta t}$$

If you look carefully at this equation, you can find a Δv/Δt, which is, by definition, acceleration. Let's replace Δv/Δt with acceleration a in the equation.

$$J = \Delta p = ma\Delta t$$

One last step... perhaps you can see it already. On the right-hand side of this equation, we have ma∆t. Utilizing Newton's 2nd Law, we can replace the product of mass and acceleration with force F, giving the final form of the equation, often-times referred to as the Impulse-Momentum Theorem:

$$J = \Delta p = F\Delta t$$

This equation, which can be found on the Regents Physics Reference Table, relates impulse to change in momentum to force applied over a time interval. To summarize, when an unbalanced force acts on an object for a period of time, a change in momentum is produced, known as an impulse.

6.11 Q: A tow-truck applies a force of 2000N on a 2000-kg car for a period of 3 seconds.

(A) What is the magnitude of the change in the car's momentum?

(B) If the car starts at rest, what will be its speed after 3 seconds?

6.11 A: (A) $\Delta p = F\Delta t = (2000N)(3s) = 6000N \bullet s$

(B) $\Delta p = p_f - p_i = mv_f - mv_i$

$$v_f = \frac{p + mv_i}{m} = \frac{6000N \bullet s + 0}{2000kg} = 3\,^m\!/_s$$

6.12 Q: A 2-kilogram body is initially traveling at a velocity of 40 meters per second east. If a constant force of 10 newtons due east is applied to the body for 5 seconds, the final speed of the body is

(1) 15 m/s
(2) 25 m/s
(3) 65 m/s
(4) 130 m/s

6.12 A: (3) $Ft = \Delta p = m\Delta v$

$$\Delta v = v_f - v_i = \frac{Ft}{m}$$

$$v_f = \frac{(10N)(5s)}{2kg} + 40\,^m\!/_s = 65\,^m\!/_s$$

6.13 Q: A motorcycle being driven on a dirt path hits a rock. Its 60-kilogram cyclist is projected over the handlebars at 20 meters per second into a haystack. If the cyclist is brought to rest in 0.50 seconds, the magnitude of the average force exerted on the cyclist by the haystack is

(1) 6.0×10^1 N

(2) 5.9×10^2 N

(3) 1.2×10^3 N

(4) 2.4×10^3 N

6.13 A: (4) $Ft = \Delta p = m\Delta v = m(v_f - v_i)$

$$F = \frac{m(v_f - v_i)}{t} = \frac{(60kg)(0 - 20 \,{}^m\!/_s)}{0.5s} = -2400N$$

6.14 Q: The instant before a batter hits a 0.14-kilogram baseball, the velocity of the ball is 45 meters per second west. The instant after the batter hits the ball, the ball's velocity is 35 meters per second east. The bat and ball are in contact for 1.0×10^{-2} second. Calculate the magnitude of the average force the bat exerts on the ball while they are in contact.

6.14 A: $Ft = \Delta p = m\Delta v$

$$F = \frac{m\Delta v}{t} = \frac{m(v_f - v_i)}{t}$$

$$F = \frac{(0.14kg)(35\,{}^m\!/_s - 45\,{}^m\!/_s)}{1 \times 10^{-2}s} = 1120N$$

6.15 Q: In an automobile collision, a 44-kilogram passenger moving at 15 meters per second is brought to rest by an air bag during a 0.10-second time interval. What is the magnitude of the average force exerted on the passenger during this time?

(1) 440 N

(2) 660 N

(3) 4400 N

(4) 6600 N

6.15 A: (4) $Ft = \Delta p$

$$F = \frac{\Delta p}{t} = \frac{p_f - p_i}{t}$$

$$F = \frac{0 - (44kg)(15\,{}^m\!/_s)}{0.1s} = -6600N$$

Non-Constant Forces

But not all forces are constant. What do you do if a changing force is applied for a period of time? In that case, we can make a graph of the force applied on the y-axis vs. time on the x-axis. The area under the Force-Time curve is the impulse, or change in momentum.

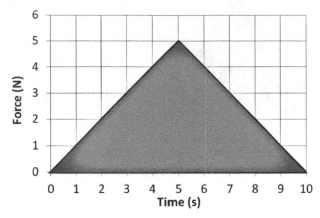

For the case of the sample graph at the right, we could determine the impulse applied by calculating the area of the triangle under the curve. In this case:

$$J = Area_{triangle} = \tfrac{1}{2}bh$$
$$J = \tfrac{1}{2}(10s)(5N) = 25N \bullet s$$

Conservation of Momentum

Now that we've talked about momentum in an isolated system, where no external forces act, we can state that momentum is always conserved. Put more simply, in any closed system, the total momentum of the system remains constant.

In the case of a collision or explosion (an event), if you add up the individual momentum vectors of all of the objects before the event, you'll find that they are equal to the sum of the momentum vectors of the objects after the event. Written mathematically, the law of conservation of momentum states:

$$P_{initial} = P_{final}$$

This is a direct outcome of Newton's 3rd Law.

In analyzing collisions and explosions, a momentum table can be a powerful tool for problem solving. To create a momentum table, follow these basic steps:

1. Identify all objects in the system. List them vertically down the left-hand column.
2. Determine the momenta of the objects before the event. Use

variables for any unknowns.

3. Determine the momenta of the objects after the event. Use variables for any unknowns.
4. Add up all the momenta from before the event and set them equal to the momenta after the event.
5. Solve your resulting equation for any unknowns.

A **collision** is an event in which two or more objects approach and interact strongly for a brief period of time. Let's look at how our problem-solving strategy can be applied to a simple collision:

6.16 Q: A 2000-kg car traveling at 20 m/s collides with a 1000-kg car at rest at a stop sign. If the 2000-kg car has a velocity of 6.67 m/s after the collision, find the velocity of the 1000-kg car after the collision.

6.16 A: Let's call the 2000-kg car Car A, and the 1000-kg car Car B. We can then create a momentum table as shown below:

Objects	Momentum Before (kg·m/s)	Momentum After (kg·m/s)
Car A	2000×20=40,000	2000×6.67=13,340
Car B	1000×0=0	1000×v_B=1000v_B
Total	40,000	13,340+1000v_B

Because momentum is conserved in any closed system, the total momentum before the event must be equal to the total momentum after the event.

$$40,000 = 13,340 + 1000v_B$$

$$v_B = \frac{40,000 - 13,340}{1000} = 26.7\,m\!/\!s$$

Not all problems are quite so simple, but problem solving steps remain consistent.

6.17 Q: On a snow-covered road, a car with a mass of 1.1×10^3 kilograms collides head-on with a van having a mass of 2.5×10^3 kilograms traveling at 8 meters per second. As a result of the collision, the vehicles lock together and immediately come to rest. Calculate the speed of the car immediately before the collision. [Neglect friction.]

6.17A: Define the car's initial velocity as positive and the van's initial velocity as negative. After the collision, the two objects join and become one, therefore we combine them in the momentum table.

Objects	Momentum Before (kg·m/s)	Momentum After (kg·m/s)
Car	$1100 \times v_{car} = 1100v_{car}$	0
Van	$2500 \times -8 = -20,000$	
Total	$-20,000 + 1100v_{car}$	0

$$-20000 + 1100v_{car} = 0$$

$$v_{car} = \frac{20000}{1100} = 18.2 \, ^m/_s$$

6.18 Q: A 70-kilogram hockey player skating east on an ice rink is hit by a 0.1-kilogram hockey puck moving toward the west. The puck exerts a 50-newton force toward the west on the player. Determine the magnitude of the force that the player exerts on the puck during this collision.

6.18 A: The player exerts a force of 50 newtons toward the east on the puck due to Newton's 3rd Law.

6.19 Q: The diagram below represents two masses before and after they collide. Before the collision, mass m_A is moving to the right with speed v, and mass m_B is at rest. Upon collision, the two masses stick together.

Before Collision

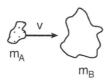

m_A

m_B

After Collision

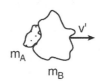

m_A

m_B

(1) $\dfrac{m_A + m_B v}{m_A}$

(2) $\dfrac{m_A + m_B}{m_A v}$

(3) $\dfrac{m_B v}{m_A + m_B}$

(4) $\dfrac{m_A v}{m_A + m_B}$

Which expression represents the speed, v′, of the masses after the collision? [Assume no outside forces are acting on m_A or m_B.]

6.19 A: Use the momentum table to set up an equation utilizing conservation of momentum, then solve for the final velocity of the combined mass, labeled v′.

Objects	Momentum Before (kg·m/s)	Momentum After (kg·m/s)
Mass A	m_Av	$(m_A+m_B)v′$
Mass B	0	
Total	m_Av	$(m_A+m_B)v′$

(4) $m_Av = (m_A + m_B)v'$

$$v' = \frac{m_Av}{m_A + m_B}$$

Let's take a look at another example which emphasizes the vector nature of momentum while examining an explosion. In physics terms, an **explosion** results when an object is broken up into two or more fragments.

6.20 Q: A 4-kilogram rifle fires a 20-gram bullet with a velocity of 300 m/s. Find the recoil velocity of the rifle.

6.20 A: Once again, we can use a momentum table to organize our problem-solving. To fill out the table, we must realize that the initial momentum of the system is 0, and we can consider the rifle and bullet as a single system with a mass of 4.02 kg:

Objects	Momentum Before (kg·m/s)	Momentum After (kg·m/s)
Rifle	0	$4×v_{recoil}$
Bullet		$(.020)(300)=6$
Total	0	$6+4×v_{recoil}$

Due to conservation of momentum, we can again state that the total momentum before must equal the total momentum after, or $0=4v_{recoil}+6$. Solving for the recoil velocity of the rifle, we find:

$$0 = 4v_{recoil} + 6$$

$$v_{recoil} = \frac{-6}{4} = -1.5\,m/s$$

The negative recoil velocity indicates the direction of the rifle's velocity. If the bullet traveled forward at 300 m/s, the rifle must travel in the opposite direction.

6.21 Q: The diagram below shows two carts that were initially at rest on a horizontal, frictionless surface being pushed apart when a compressed spring attached to one of the carts is released. Cart A has a mass of 3.0 kilograms and cart B has a mass of 5.0 kilograms.

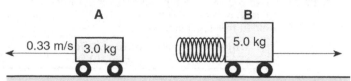

If the speed of cart A is 0.33 meter per second after the spring is released, what is the approximate speed of cart B after the spring is released?

(1) 0.12 m/s

(2) 0.20 m/s

(3) 0.33 m/s

(4) 0.55 m/s

6.21 A: Define the positive direction toward the right of the page.

Objects	Momentum Before (kg·m/s)	Momentum After (kg·m/s)
Cart A	0	3×-0.33=-1
Cart B	0	5×v$_B$
Total	0	5v$_B$-1

(2) $0 = 5v_B - 1$

$$v_B = \frac{1}{5} = 0.2 \, ^m/_s$$

6.22 Q: A woman with horizontal velocity v$_1$ jumps off a dock into a stationary boat. After landing in the boat, the woman and the boat move with velocity v$_2$. Compared to velocity v$_1$, velocity v$_2$ has

(1) the same magnitude and the same direction

(2) the same magnitude and opposite direction

(3) smaller magnitude and the same direction

(4) larger magnitude and the same direction

6.22 A: (3) smaller magnitude and the same direction due to the law of conservation of momentum.

You can find more practice problems on the APlusPhysics website at: http://www.aplusphysics.com/regents.

Chapter 7: Work, Energy & Power

"Ambition is like a vector; it needs magnitude and direction.
Otherwise, it's just energy."

— Grace Lindsay

Objectives

1. Define work and calculate the work done by a force.
2. Calculate the kinetic energy of a moving object.
3. Determine the gravitational potential energy of a system.
4. Calculate the power of a system.
5. Apply conservation of energy to analyze energy transitions and transformations in a system.
6. Analyze the relationship between work done on or by a system, and the energy gained or lost by that system.
7. Use Hooke's Law to determine the elastic force on an object.
8. Calculate a system's elastic potential energy.

Work, energy and power are highly inter-related concepts that come up regularly in everyday life. You do work on an object when you move it. The rate at which you do the work is your power output. When you do work on an object, you transfer energy from one object to another. In this chapter we'll explore how energy is transferred and transformed, how doing work on an object changes its energy, and how quickly work can be done.

Work

Sometimes we work hard. Sometimes we're slackers. But, right now, are you doing work? And what do we mean by work? In physics terms, **work** is the process of moving an object by applying a force.

I'm sure you can think up countless examples of work being done, but a few that spring to my mind include pushing a snowblower to clear the driveway, pulling a sled up a hill with a rope, stacking boxes of books from the floor onto a shelf, and throwing a baseball from the pitcher's mound to home plate.

Let's take a look at a few scenarios and investigate what work is being done.

In the first scenario, a monkey in a jet pack blasts through the atmosphere, accelerating to higher and higher speeds. In this case, the jet pack is applying a force causing it to move. But what is doing the work? Hot expanding gases are pushed backward out of the jet pack. Using Newton's 3rd Law, we observe the reactionary force of the gas pushing the jet pack forward, causing a displacement. Therefore, the expanding exhaust gas is doing work on the jet pack.

In the second scenario, a girl struggles to push her stalled car, but can't make it move. Even though she's expending significant effort, no work is being done on the car because it isn't moving.

In our final scenario, a child in a ghost costume carries a bag of Halloween candy across the yard. In this situation, the child applies a force upward on the bag, but the bag moves horizontally. From this perspective, the forces of the child's arms on the bag don't cause the displacement, therefore no work is being done by the child.

Mathematically, work can be expressed by the following equation:

$$W = Fd$$

W is the work done, **F** is the force applied in Newtons, and **d** is the object's displacement in meters.

The units of work can be found by performing unit analysis on the work formula. If work is force multiplied by distance, the units must be the units of force multiplied by the units of distance, or newtons multiplied by meters. A newton-meter is also known as a Joule (J).

It's important to note that when using this equation, only the force applied in the direction of the object's displacement counts! This means that if the force and displacement vectors aren't in exactly the same direction, you need to take the component of force in the direction of the object's displacement. To do this, line up the force and displacement vectors tail-to-tail and measure the angle between them.
Since this component of force can be calculated by multiplying the force by the cosine of the angle between the force and displacement vectors, we can re-write the work equation as:

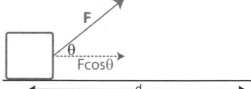

$$W = F\cos\theta \bullet d = Fd\cos\theta$$

7.01 Q: An appliance salesman pushes a refrigerator 2 meters across the floor by applying a force of 200N. Find the work done.

7.01 A: Since the force and displacement are in the same direction, the angle between them is 0.

$$W = Fd\cos\theta = (200N)(2m)\cos 0° = 400J$$

7.02 Q: A friend's car is stuck on the ice. You push down on the car to provide more friction for the tires (by way of increasing the normal force), allowing the car's tires to propel it forward 5m onto less slippery ground. How much work did you do?

7.02 A: You applied a downward force, yet the car's displacement was sideways. Therefore, the angle between the force and displacement vectors is 90°.

$$W = Fd\cos\theta = Fd\cos 90° = 0$$

7.03 Q: You push a crate up a ramp with a force of 10N. Despite your pushing, however, the crate slides down the ramp a distance of 4m. How much work did you do?

7.03 A: Since the direction of the force you applied is opposite the direction of the crate's displacement, the angle between the two vectors is 180°.

$$W = Fd\cos\theta = (10N)(4m)\cos 180° = -40J$$

7.04 Q: How much work is done in lifting an 8-kg box from the floor to a height of 2m above the floor?

7.04 A: It's easy to see the displacement is 2m, and the force must be applied in the direction of the displacement, but what is the force? To lift the box you must match and overcome the force of gravity on the box. Therefore, the force applied is equal to the gravitational force, or weight, of the box, mg=(8kg)(9.81m/s²)=78.5N.

$$W = Fd\cos\theta = (78.5N)(2m)\cos 0° = 157J$$

7.05 Q: Barry and Sidney pull a 30-kg wagon with a force of 500N a distance of 20m. The force acts at a 30° angle to the horizontal. Calculate the work done.

7.05 A: $W = Fd\cos\theta = (500N)(20m)\cos 30° = 8660J$

7.06 Q: The work done in lifting an apple one meter near Earth's surface is approximately

(1) 1 J

(2) 0.01 J

(3) 100 J

(4) 1000 J

7.06 A: (1) The trick in this problem is recalling the approximate weight of an apple. With an "order-of-magnitude" estimate, we can say an apple has a mass of 0.1 kg, or a weight of 1 N. Given this information, the work done is:

$$W = Fd\cos\theta = (1N)(1m)\cos 0° = 1J$$

7.07 Q: As shown in the diagram, a child applies a constant 20-newton force along the handle of a wagon which makes a 25° angle with the horizontal.

How much work does the child do in moving the wagon a horizontal distance of 4.0 meters?

(1) 5.0 J

(2) 34 J

(3) 73 J

(4) 80. J

7.07 A: (4) $W = Fd\cos\theta = (20N)(4m)\cos(25°) = 73J$

Force vs. Displacement Graphs

The area under a force vs. displacement graph is the work done by the force. Consider the situation of a block being pulled across a table with a constant force of 5 Newtons over a displacement of 5 meters, then the force gradually tapers off over the next 5 meters.

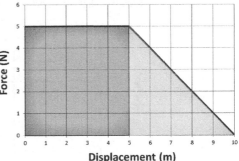

The work done by the force moving the block can be calculated by taking the area under the force vs. displacement graph (a combination of a rectangle and triangle) as follows:

$$Work = Area_{rectangle} + Area_{triangle}$$
$$Work = lw + \frac{1}{2}bh$$
$$Work = (5m)(5N) + \frac{1}{2}(5m)(5N)$$
$$Work = 37.5J$$

7.08 Q: A boy pushes his wagon at constant speed along a level sidewalk. The graph below represents the relationship between the horizontal force exerted by the boy and the distance the wagon moves.

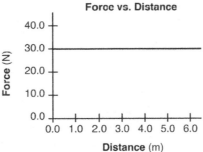

What is the total work done by the boy in pushing the wagon 4.0 meters?

(1) 5.0 J

(2) 7.5 J

(3) 120 J

(4) 180 J

7.08 A: (3) 120 J

$$Work = Area_{rectangle} = lw = (4m)(30N) = 120J$$

7.09 Q: A box is wheeled to the right with a varying horizontal force. The graph below represents the relationship between the applied force and the distance the box moves.

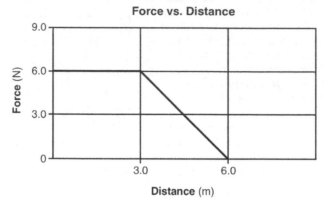

Force vs. Distance

What is the total work done in moving the box 6 meters?

(1) 9.0 J

(2) 18 J

(3) 27 J

(4) 36 J

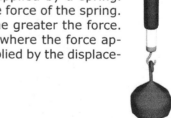

7.09 A: (3) $Work = Area_{rectangle} + Area_{triangle}$

$Work = lw + \frac{1}{2}bh$

$Work = (3m)(6N) + \frac{1}{2}(3m)(6N)$

$Work = 27J$

Hooke's Law

An interesting application of work combined with the Force and Displacement graph is examining the force applied by a spring. The more you stretch a spring, the greater the force of the spring. Similarly, the more you compress a spring, the greater the force. This can be modeled as a linear relationship, where the force applied by the spring is equal to a constant multiplied by the displacement of the spring.

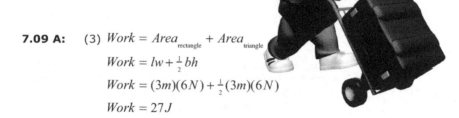

$$F_s = kx$$

F_s is the force of the spring in newtons, x is the displacement of the spring from its equilibrium (or rest) position, in meters, and k is the spring constant, which tells you how stiff or powerful a spring is, in newtons per meter. The larger the spring constant, k, the more force the spring applies per amount of displacement.

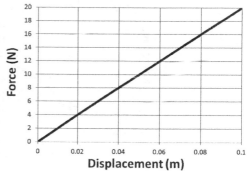

You can determine the spring constant of a spring by making a graph of the force from a spring on the y-axis, and placing the displacement of the spring from its equilibrium, or rest position, on the x-axis. The slope of the graph will give you the spring constant. For the case of the spring depicted in the graph at right, we can find the spring constant as follows:

$$k = Slope = \frac{rise}{run} = \frac{20N - 0N}{0.1m - 0m} = 200\,^N/_m$$

You must have done work to compress or stretch the spring, since you applied a force and caused a displacement. You can find the work done in stretching or compressing a spring by taking the area under the graph. For the spring shown, to displace the spring 0.1m, you can find the work done as shown below:

$$Work = Area_{tri} = \tfrac{1}{2}bh = \tfrac{1}{2}(0.1m)(20N) = 1J$$

7.10 Q: In an experiment, a student applied various forces to a spring and measured the spring's corresponding elongation. The table below shows his data.

Force (newtons)	Elongation (meters)
0	0
1.0	0.30
3.0	0.67
4.0	1.00
5.0	1.30
6.0	1.50

Plot force versus elongation and draw the best-fit line. Then, using your graph, calculate the spring constant of the spring. [Show all work, including the equation and substitution with units.]

7.10 A:

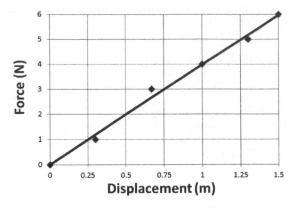

$$k = Slope = \frac{rise}{run} = \frac{\Delta F}{\Delta x} = \frac{6N - 0N}{1.5m - 0m} = 4\,{}^N\!/_m$$

7.11 Q: In a laboratory investigation, a student applied various down-ward forces to a vertical spring. The applied forces and the cor-responding elongations of the spring from its equilibrium position are recorded in the data table.

Construct a graph, marking an appropriate scale on the axis la-beled "Force (N)." Plot the data points for force versus elongation. Draw the best-fit line or curve. Then, using your graph, calcu-late the spring constant of this spring. [Show all work, including the equation and substitution with units.]

Force (newtons)	Elongation (meters)
0	0
0.5	0.010
1.0	0.018
1.5	0.027
2.0	0.035
2.5	0.046

7.11 A:

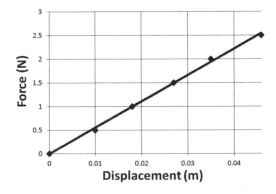

$$k = Slope = \frac{rise}{run} = \frac{\Delta F}{\Delta x} = \frac{2.5N - 0.8N}{0.046m - 0.015m} = 55\,{}^N\!/_m$$

7.12 Q: A 10-newton force compresses a spring 0.25 meter from its equilibrium position. Calculate the spring constant of this spring.

7.12 A: $F_s = kx$

$$k = \frac{F_s}{x} = \frac{10N}{0.25m} = 40 \, {}^{N}\!/_{m}$$

Power

Power is a term used quite regularly in all aspects of life. We talk about how powerful the new boat motor is, the power of positive thinking, and even the power company's latest bill. All of these uses of the term power relate to how much work can be done in some amount of time.

In physics, work is the process of moving an object by applying a force. The rate at which the force does work is known as **power** (P). The units of power are the units of work divided by time, or Joules per second, known as a watt (W).

$$P = \frac{W}{t}$$

Since power is the rate at which work is done, it is possible to have the same amount of work done but with a different supplied power, if the time is different.

7.13 Q: Rob and Peter move a sofa 3 meters across the floor by applying a combined force of 200N horizontally. If it takes them 6 seconds to move the sofa, what amount of power did they supply?

7.13 A: $P = \dfrac{W}{t} = \dfrac{Fd\cos\theta}{t} = \dfrac{(200N)(3m)}{6s} = 100W$

7.14 Q: Kevin then pushes the same sofa 3 meters across the floor by applying a force of 200N. Kevin, however, takes 12 seconds to push the sofa. What amount of power did Kevin supply?

7.14 A: $P = \dfrac{W}{t} = \dfrac{Fd\cos\theta}{t} = \dfrac{(200N)(3m)}{12s} = 50W$

As you can see, although Kevin did the same amount of work as Rob and Peter in pushing the sofa (600J), Rob and Peter supplied twice the power of Kevin because they did the same work in half the time!

There's more to the story, however. Since power is defined as work over time, and because work is equal to force (in the direction of the displacement) multiplied by displacement, we can replace work in the equation with **F×d**:

$$P = \frac{W}{t} = \frac{Fd}{t}$$

Looking carefully at this equation, you can observe a displacement divided by time. Since displacement divided by time is the definition of average velocity, we can replace d/t with v in the equation to obtain:

$$P = \frac{W}{t} = \frac{Fd}{t} = F\overline{v}$$

So, not only is power equal to work done divided by the time required, it's also equal to the force applied (in the direction of the displacement) multiplied by the average velocity of the object.

7.15 Q: Motor A lifts a 5000N steel crossbar upward at a constant 2 m/s. Motor B lifts a 4000N steel support upward at a constant 3 m/s. Which motor is supplying more power?

7.15 A: Motor B supplies more power than Motor A.

$P_{MotorA} = F\overline{v} = (5000N)(2\,^m/_s) = 10000W$

$P_{MotorB} = F\overline{v} = (4000N)(3\,^m/_s) = 12000W$

7.16 Q: A 70-kilogram cyclist develops 210 watts of power while pedaling at a constant velocity of 7 meters per second east. What average force is exerted eastward on the bicycle to maintain this constant speed?

(1) 490 N

(2) 30 N

(3) 3.0 N

(4) 0 N

Chapter 7: Work, Energy & Power

7.16 A: (2) $P = \overline{F}\overline{v}$

$$F = \frac{P}{\overline{v}} = \frac{210W}{7\,m/_s} = 30N$$

7.17 Q: Alien A lifts a 500-newton child from the floor to a height of 0.40 meters in 2 seconds. Alien B lifts a 400-newton student from the floor to a height of 0.50 meters in 1 second. Compared to Alien A, Alien B does

(1) the same work but develops more power

(2) the same work but develops less power

(3) more work but develops less power

(4) less work but develops more power

7.17 A: (1) the same work but develops more power.

7.18 Q: A 110-kilogram bodybuilder and his 55-kilogram friend run up identical flights of stairs. The bodybuilder reaches the top in 4.0 seconds while his friend takes 2.0 seconds. Compared to the power developed by the bodybuilder while running up the stairs, the power developed by his friend is

(1) the same

(2) twice as much

(3) half as much

(4) four times as much

7.18 A: (1) the same.

Energy

We've all had days where we've had varying amounts of energy. You've gotten up in the morning, had to drag yourself out of bed, force yourself to get ready to school, and once you finally get to class, you don't have the energy to do much work. Other days, when you've had more energy, you may have woken up before the alarm clock, hustled to get ready for the day while a bunch of thoughts jump around in your head, and hurried on to begin your activities. Then, throughout the day, the more work you do, the more energy you lose... What's the difference in these days?

In physics, **energy** is the ability or capacity to do work. And as we've mentioned previously, work is the process of moving an object. So, if we combine our definitions, energy is the ability or capacity to move an object. So far we've examined kinetic energy, or energy of motion, and therefore kinetic energy must be the ability or capacity of a moving object to move another object! Mathematically, kinetic energy is calculated using the formula:

$$KE = \tfrac{1}{2}mv^2$$

Of course, there are more types of energy than just kinetic. Energy comes in many forms, which we can classify as kinetic (energy of motion) or potential (stored) to various degrees (This includes solar energy, thermal energy, gravitational potential energy, nuclear energy, chemical potential energy, sound energy, electrical energy, elastic potential energy, light energy, and so on). In all cases, energy can be transformed from one type to another and you can transfer energy from one object to another by doing work.

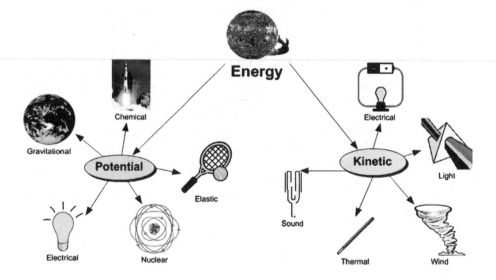

The units of energy are the same as the units of work, joules (J). By dimensional analysis, we observe that the units of KE (kg·m²/s²) must be equal to the units of work (N·m):

$$\frac{kg \bullet m^2}{s^2} = N \bullet m = J$$

7.19 Q: Which is an SI unit for work done on an object?

(1) $\dfrac{kg \bullet m^2}{s^2}$ (3) $\dfrac{kg \bullet m}{s}$

$$(2) \ \frac{kg \bullet m^2}{s} \qquad\qquad (4) \ \frac{kg \bullet m}{s^2}$$

7.19 A: (1) is equivalent to a newton-meter, also known as a Joule.

Gravitational Potential Energy

Potential energy is energy an object possesses due to its position or condition. **Gravitational potential energy** is the energy an object possesses because of its position in a gravitational field (height).

Assume we have a 10 kg box on the floor. Let's arbitrarily call its current potential energy zero, just to give us a reference point. If we do work to lift the box one meter off the floor, we need to overcome the force of gravity on the box (its weight) over a distance of one meter. Therefore, the work we do on the box can be obtained from:

$$W = Fd = (mg)h = (10kg)(9.81 \, ^m/_{s^2})(1m) = 98.1J$$

So, to raise the box to a height of 1m, we must do 98.1 Joules of work on the box. The work done in lifting the box is equal to the change in the potential energy of the box, so the box's gravitational potential energy must be 98.1J.

When we performed work on the box, we transferred some of our stored energy to the box. Along the way, it just so happens that we derived the formula for the gravitational potential energy of an object. The change in the object's potential energy, ΔPE, is equal to the force of gravity on the box multiplied by its change in height, mgΔh. This formula can be found on the reference table:

$$\Delta PE = mg\Delta h$$

This formula can be used to solve a variety of problems involving the potential energy of an object.

7.20 Q: The diagram below represents a 155-newton box on a ramp. Applied force F causes the box to slide from point A to point B.

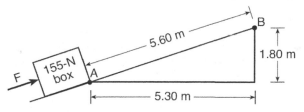

What is the total amount of gravitational potential energy gained by the box?

(1) 28.4 J

(2) 279 J

(3) 868 J

(4) 2740 J

7.20 A: (2) $\Delta PE = mg\Delta h = (155N)(1.8m) = 279J$

7.21 Q: Which situation describes a system with decreasing gravitational potential energy?

(1) a girl stretching a horizontal spring

(2) a bicyclist riding up a steep hill

(3) a rocket rising vertically from Earth

(4) a boy jumping down from a tree limb

7.21 A: (4) The boy's height above ground is decreasing, so his gravitational PE is decreasing.

7.22 Q: A car travels at constant speed v up a hill from point A to point B, as shown in the diagram below.

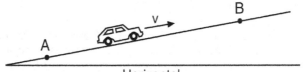

As the car travels from A to B, its gravitational potential energy

(1) increases and its kinetic energy decreases

(2) increases and its kinetic energy remains the same

(3) remains the same and its kinetic energy decreases

(4) remains the same and its kinetic energy remains the same

7.22 A: (2) The car's height above ground increases so gravitational potential energy increases, and velocity remains constant, so kinetic energy remains the same.

7.23 Q: An object is thrown vertically upward. Which pair of graphs best represents the object's kinetic energy and gravitational potential energy as functions of its displacement while it rises?

Chapter 7: Work, Energy & Power

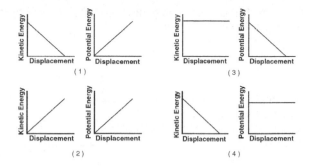

(1) (2) (3) (4)

7.23 A: (1) shows the object's kinetic energy decreasing as it slows down on its way upward, while its potential energy increases as its height increases.

7.24 Q: While riding a chairlift, a 55-kilogram snowboarder is raised a vertical distance of 370 meters. What is the total change in the snowboarder's gravitational potential energy?

(1) 5.4×10^1 J

(2) 5.4×10^2 J

(3) 2.0×10^4 J

(4) 2.0×10^5 J

7.24 A: (4) 2.0×10^5 J

$$\Delta PE = mg\Delta h = (55kg)(9.81\tfrac{m}{s^2})(370m) = 2 \times 10^5 J$$

Elastic Potential Energy

Another form of potential energy involves the stored energy an object possesses due to its position in a stressed elastic system. An object at the end of a compressed spring, for example, has **elastic potential energy**. When the spring is released, the elastic potential energy of the spring will do work on the object, moving the object and transferring the energy of the spring into kinetic energy of the object. Other examples of elastic potential energy include tennis rackets, rubber bands, bows (as in bows and arrows), trampolines, bouncy balls, and even pole-vaulting poles.

The most common problems involving elastic potential energy in Regents Physics involve the energy stored in a spring. As we learned in the previous topic on work, the force needed to compress or stretch a spring from its equilibrium position increases linearly. The more you stretch or compress

---

the spring, the more force it applies trying to restore itself to its equilibrium position. We called this Hooke's Law:

$$F_s = kx$$

Further, we can find the work done in compressing or stretching the spring by taking the area under a Force vs. Displacement graph for the spring.

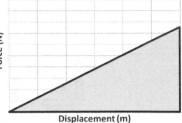

$$W = Fd = Area_{triangle} = \tfrac{1}{2}bh = \tfrac{1}{2}(x)(kx) = \tfrac{1}{2}kx^2$$

Since the work done in compressing or stretching the spring from its equilibrium position transfers energy to the spring, we can conclude that the potential energy stored in the spring must be equal to the work done to compress the spring, therefore we can write:

$$PE_s = \tfrac{1}{2}kx^2$$

7.25 Q: A spring with a spring constant of 4.0 newtons per meter is compressed by a force of 1.2 newtons. What is the total elastic potential energy stored in this compressed spring?

(1) 0.18 J

(2) 0.36 J

(3) 0.60 J

(4) 4.8 J

7.25 A: PE_s can't be calculated directly since x isn't known, but x can be found from Hooke's Law:

$$F_s = kx$$

$$x = \frac{F_s}{k} = \frac{1.2N}{4\,{}^N\!/_m} = 0.3m$$

With x known, the potential energy equation for a spring can be utilized.

$$PE_s = \tfrac{1}{2}kx^2$$

$$PE_s = \tfrac{1}{2}(4\,{}^N\!/_m)(0.3m)^2 = 0.18J$$

7.26 Q: An unstretched spring has a length of 10 centimeters. When the spring is stretched by a force of 16 newtons, its length is increased to 18 centimeters. What is the spring constant of this spring?

(1) 0.89 N/cm

(2) 2.0 N/cm

---

(3) 1.6 N/cm

(4) 1.8 N/cm

7.26 A: $F_s = kx$

$$k = \frac{F_s}{x} = \frac{16N}{8cm} = 2\,{}^{N}\!/_{cm}$$

7.27 Q: Which graph best represents the relationship between the elastic potential energy stored in a spring and its elongation from equilibrium?

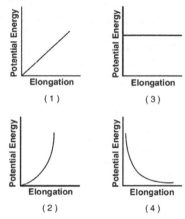

7.27 A: (2) due to the displacement^2 relationship.

7.28 Q: A pop-up toy has a mass of 0.020 kilogram and a spring constant of 150 newtons per meter. A force is applied to the toy to compress the spring 0.050 meter.

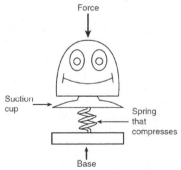

(A) Calculate the potential energy stored in the compressed spring.

(B) The toy is activated and all the compressed spring's potential energy is converted to gravitational potential energy. Calculate the maximum vertical height to which the toy is propelled.

7.28 A: (A) $PE_s = \frac{1}{2}kx^2$

$$PE_s = \frac{1}{2}(150 \, {}^N\!/_m)(0.05m)^2 = 0.1875J$$

(B) $PE_g = mgh$

$$h = \frac{PE_g}{mg} = \frac{0.1875J}{(0.02kg)(9.81 \, {}^m\!/_{s^2})} = 0.96m$$

7.29 Q: A spring with a spring constant of 80 newtons per meter is displaced 0.30 meter from its equilibrium position. The potential energy stored in the spring is

(1) 3.6 J

(2) 7.2 J

(3) 12 J

(4) 24 J

7.29 A: (1) $PE_s = \frac{1}{2}kx^2 = \frac{1}{2}(80 \, {}^N\!/_m)(0.3m)^2 = 3.6J$

Work-Energy Theorem

Of course, there are many different kinds of energy which we haven't mentioned specifically. Energy can be converted among its many different forms, such as mechanical (which is kinetic, gravitational potential, and elastic potential), electromagnetic, nuclear, and thermal (or internal) energy.

When a force does work on a system, the work done changes the system's energy. If the work done increases motion, there is an increase in the system's kinetic energy. If the work done increases the object's height, there is an increase in the system's gravitational potential energy. If the work done compresses a spring, there is an increase in the system's elastic potential energy. If the work is done against friction, however, where does the energy go? In this case, the energy isn't lost, but instead increases the rate at which molecules in the object vibrate, increasing the object's temperature, or internal energy.

The understanding that the work done on a system by an external force changes the energy of the system is known as the Work-Energy Theorem. If an external force does positive work on the system, the system's total energy increases. If, instead, the system does work, the system's total energy decreases. Put another way, you add energy to a system by doing work on it and take energy from a system when the system does the work (much like you add value to your bank account by making a deposit and take value from your account by writing a check).

This relationship is documented on the Regents Physics Reference Table by showing the formula for work as equal to the force times the displacement (F·d), as well as the change in total energy (ΔE):

$$W = Fd = \Delta E_T$$

Sources of Energy on Earth

So where does all this energy initially come from? Here on Earth, the energy we deal with everyday ultimately comes from the conversion of mass into energy, the source of the sun's energy. The sun's radiation provides an energy source for life on earth, which over the millennia has become the source of our fossil fuels. The sun's radiation also provides the thermal and light energy that heat the atmosphere and cause the winds to blow. The sun's energy evaporates water, which eventually recondenses as rain and snow, falling to the Earth's surface to create lakes and rivers, with gravitational potential energy, which we may harness in hydroelectric power plants. Nuclear power also comes from the conversion of mass into energy. Just try to find an energy source on Earth that doesn't originate with the conversion of mass into energy!

Conservation of Energy

"Energy cannot be created or destroyed... it can only be changed."

Chances are you've heard that phrase before. It's one of the most important concepts in all of physics. It doesn't mean that an object can't lose energy or gain energy. What it means is that energy can be changed into different forms, and transferred from system to system, but it never magically disappears or reappears.

Following up on our bank account analogy, if you have a certain amount of money in your bank account and then you spend some money, your bank account balance decreases. Your money wasn't lost, however, it was transferred to another system. It may even change forms... perhaps you purchased an item from another country. Your money is no longer in the form of dollars and cents, but is instead now part of another system in another currency.

There are some issues with our money analogy, however. If the total money supply in a country is low, a government can print more currency. In the world of physics, however, the total amount of energy throughout the universe is fixed. In other words, it cannot be replenished. Alternately, governments can collect and destroy currency -- in the world of physics, we can never truly destroy energy. The understanding that the total amount of energy in the universe remains fixed is known as the law of conservation of energy.

Mechanical energy is the sum of an object's kinetic energy as well as its gravitational potential and elastic potential energies. Non-mechanical energy forms include chemical potential, nuclear, and thermal.

Total energy is always conserved in any system, which is the law of conservation of energy. By confining ourselves to just the mechanical forms of energy, however, if we neglect the effects of friction we can also state that total mechanical energy is constant in any system.

Let's take the example of an F/A-18 Hornet jet fighter with a mass of 20,000 kilograms flying at an altitude of 10,000 meters above the surface of the earth with a velocity of 250 m/s. In this scenario, we can calculate the total mechanical energy of the jet fighter as follows:

$$E_T = PE_g + KE = mgh + \tfrac{1}{2}mv^2$$
$$E_T = (20000 kg)(9.81 \, \tfrac{m}{s^2})(10000m) + \tfrac{1}{2}(20000 kg)(250 \, \tfrac{m}{s})^2$$
$$E_T = 2.59 \times 10^9 J$$

Now, let's assume the Hornet dives down to an altitude of 2,000 meters above the surface of the Earth. Total mechanical energy remains constant, and the gravitational potential energy of the fighter decreases, therefore the kinetic energy of the fighter must increase. The fighter's velocity goes up as a result of flying closer to the Earth! For this reason, a key concept in successful dogfighting taught to military pilots is that of energy conservation!

We can even calculate the new velocity of the fighter jet since we know its new height and its total mechanical energy must remain constant. Solving for velocity, we find that the Hornet has almost doubled its speed by "trading in" 8000 meters of altitude!

$$E_T = PE_g + KE = mgh + \tfrac{1}{2}mv^2$$

$$\tfrac{1}{2}mv^2 = E_T - mgh$$

$$v = \sqrt{\frac{2(E_T - mgh)}{m}}$$

$$v = \sqrt{\frac{2(2.59 \times 10^9 \, J - (20000kg)(9.81 \, ^m\!/_{s^2})(2000m))}{20000kg}}$$

$$v = 469 \, ^m\!/_s$$

If instead we had been told that some of the mechanical energy of the jet was lost to air resistance (friction), we could also account for that by stating that the total mechanical energy of the system is equal to the gravitational potential energy, the kinetic energy, and the change in internal energy of the system (Q). This leads to the conservation of mechanical energy formula given on the Regents Reference Table:

$$E_T = PE + KE + Q$$

Let's take another look at free fall, only this time, let's analyze our falling object using the law of conservation of energy and compare it to our analysis using the kinematics equations we studied previously.

The problem: An object falls from a height of 10m above the ground. Neglecting air resistance, find its velocity the moment before the object strikes the ground.

Conservation of Energy Approach: The energy of the object at its highest point must equal the energy of the object at its lowest point, therefore:

$$E_{top} = E_{bottom}$$

$$PE_{top} = KE_{bottom}$$

$$mgh_{top} = \tfrac{1}{2}mv^2_{bottom}$$

$$v_{bottom} = \sqrt{2gh} = \sqrt{2(9.81 \, ^m\!/_{s^2})(10m)} = 14 \, ^m\!/_s$$

Kinematics Approach: For an object in free fall, its initial velocity must be zero, its displacement is 10 meters, and the acceleration due to gravity on the surface of the Earth is 9.81 m/s². Choosing down as the positive direction:

Variable	Value
v_i	0 m/s
v_f	FIND
d	10 m
a	9.81 m/s²
t	?

$$v_f^2 = v_i^2 + 2ad$$

$$v_f = \sqrt{v_i^2 + 2ad} = \sqrt{2(9.81\,^m/_{s^2})(10m)} = 14\,^m/_s$$

As you can see, we reach the same conclusion regardless of approach!

7.30 Q: The diagram below shows a toy cart possessing 16 joules of kinetic energy traveling on a frictionless, horizontal surface toward a horizontal spring.

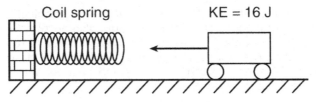

Coil spring   KE = 16 J

Frictionless, horizontal surface

If the cart comes to rest after compressing the spring a distance of 1.0 meter, what is the spring constant of the spring?

(1) 32 N/m

(2) 16 N/m

(3) 8.0 N/m

(4) 4.0 N/m

7.30 A: (1) $KE = PE_s = \frac{1}{2}kx^2$

$$k = \frac{2KE}{x^2} = \frac{2(16J)}{(1m)^2} = 32\,^N/_m$$

7.31 Q: A child does 0.20 joules of work to compress the spring in a pop-up toy. If the mass of the toy is 0.010 kilograms, what is the maximum vertical height that the toy can reach after the spring is released?

(1) 20 m

(2) 2.0 m

(3) 0.20 m

(4) 0.020 m

7.31 A: (2) The potential energy in the compressed spring must be equal to the gravitational potential energy of the toy at its maximum vertical height.

$$PE_s = PE_g = mgh$$

$$h = \frac{PE_s}{mg} = \frac{0.2J}{(0.01kg)(9.81\,{}^m/_{s^2})} = 2m$$

7.32 Q: A lawyer knocks her folder of mass m off her desk of height h. What is the speed of the folder upon striking the floor?

(1) $\sqrt{(2gh)}$

(2) 2gh

(3) mgh

(4) mh

7.32 A: (1) The folder's initial gravitational energy becomes its kinetic energy right before striking the floor.

$$PE_{desk} = KE_{floor}$$

$$mgh = \tfrac{1}{2}mv^2$$

$$v = \sqrt{2gh}$$

7.33 Q: A 65-kilogram pole vaulter wishes to vault to a height of 5.5 meters.

(A) Calculate the minimum amount of kinetic energy the vaulter needs to reach this height if air friction is neglected and all the vaulting energy is derived from kinetic energy.

(B) Calculate the speed the vaulter must attain to have the necessary kinetic energy.

7.33 A: (A) $KE = \Delta PE = mg\Delta h$

$$KE = (65kg)(9.81\,{}^m/_{s^2})(5.5m) = 3500J$$

(B) $KE = \tfrac{1}{2}mv^2$

$$v = \sqrt{\frac{2KE}{m}} = \sqrt{\frac{2(3500J)}{65kg}} = 10\,{}^m/_s$$

7.34 Q: A car, initially traveling at 30 meters per second, slows uniformly as it skids to a stop after the brakes are applied. Sketch a graph showing the relationship between the kinetic energy of the car as it is being brought to a stop and the work done by friction in stopping the car.

7.34 A:

7.35 Q: The work done in accelerating an object along a frictionless horizontal surface is equal to the change in the object's

(1) momentum

(2) velocity

(3) potential energy

(4) kinetic energy

7.35 A: (4) Due to the Work-Energy Theorem.

7.36 Q: A 2-kilogram block sliding down a ramp from a height of 3 meters above the ground reaches the ground with a kinetic energy of 50 joules. The total work done by friction on the block as it slides down the ramp is approximately

(1) 6 J

(2) 9 J

(3) 18 J

(4) 44 J

7.36 A: (2) The box has gravitational potential energy at the top of the ramp, which is converted to kinetic energy as it slides down the ramp. Any gravitational potential energy not converted to kinetic energy must be the work done by friction on the block, converted to internal energy (heat) of the system.

$$PE_{top} = KE_{bottom} + W_{friction}$$

$$W_{friction} = PE_{top} - KE_{bottom} = mgh - KE_{bottom}$$

$$W_{friction} = (2kg)(9.81 \, \sqrt[m]{s^2})(3m) - 50J = 9J$$

You can find more practice problems on the APlusPhysics website at: http://www.aplusphysics.com/regents.

Chapter 8: Electrostatics

"Electricity is really just organized lightning."

— George Carlin

Objectives

1. Calculate the charge on an object.
2. Describe the differences between conductors and insulators.
3. Explain the difference between conduction and induction.
4. Explain how an electroscope works.
5. Solve problems using the law of conservation of charge.
6. Use Coulomb's Law to solve problems related to electrical force.
7. Recognize that objects that are charged exert forces, both attractive and repulsive.
8. Compare and contrast Newton's Law of Universal Gravitation with Coulomb's Law.
9. Define, measure, and calculate the strength of an electric field.
10. Solve problems related to charge, electric field, and forces.
11. Define and calculate electric potential energy.
12. Define and calculate potential difference.
13. Solve basic problems involving charged parallel plates.

Electricity and magnetism play a profound role in almost all aspects of our daily lives. From the moment we wake up, to the moment we go to sleep (and even while we're sleeping), applications of electricity and magnetism provide us tools, light, warmth, transportation, communication, and even entertainment. Despite its widespread use, however, there is much we're still learning every day about these phenomena!

Electric Charges

Matter is made up of atoms. Once thought to be the smallest building blocks of matter, we now know that atoms can be broken up into even smaller pieces, known as protons, electrons, and neutrons. Each atom consists of a dense core of positively charged protons and uncharged (neutral) neutrons. This core is known as the nucleus. It is surrounded by a "cloud" of much smaller, negatively charged electrons. These electrons orbit the nucleus in distinct energy levels. To move to a higher energy level, an electron must absorb energy. When an electron falls to a lower energy level, it gives off energy.

Most atoms are neutral -- that is, they have an equal number of positive and negative charges, giving a net charge of 0. In certain situations, however, an atom may gain or lose electrons. In these situations, the atom as a whole is no longer neutral and we call it an **ion**. If an atom loses one or more electrons, it has a net positive charge and is known as a positive ion. If, instead, an atom gains one or more electrons, it has a net negative charge and is therefore called a negative ion. Like charges repel each other, while opposite charges attract each other. In physics, we represent the charge on an object with the symbol q.

Charge is a fundamental measurement in physics, much as length, time, and mass are fundamental measurements. The fundamental unit of charge is the Coulomb [C], which is a very large amount of charge. Compare that to the magnitude of charge on a single proton or electron, known as an elementary charge, which is equal to 1.6×10^{-19} coulomb. It would take 6.25×10^{18} elementary charges to make up a single coulomb of charge! (Don't worry about memorizing these values, they're listed for you on the front of the reference table).

8.01 Q: An object possessing an excess of 6.0×10^6 electrons has what net charge?

8.01 A: $6 \times 10^6 \text{ electrons} \bullet \dfrac{-1.6 \times 10^{-19} C}{1 \text{ electron}} = -9.6 \times 10^{-13} C$

8.02 Q: An alpha particle consists of two protons and two neutrons. What is the charge of an alpha particle?

(1) 1.25×10^{19} C

(2) 2.00 C

(3) 6.40×10^{-19} C

(4) 3.20×10^{-19} C

8.02 A: (4) The net charge on the alpha particle is +2 elementary charges.

$$2e \bullet \frac{1.6 \times 10^{-19} C}{1e} = 3.2 \times 10^{-19} C$$

8.03 Q: If an object has a net negative charge of 4 coulombs, the object possesses

(1) 6.3×10^{18} more electrons than protons

(2) 2.5×10^{19} more electrons than protons

(3) 6.3×10^{18} more protons than electrons

(4) 2.5×10^{19} more protons than electrons

8.03 A: (2) $-4C \bullet \dfrac{1e}{1.6 \times 10^{-19} C} = -2.5 \times 10^{19} e$

8.04 Q: Which quantity of excess electric charge could be found on an object?

(1) 6.25×10^{-19} C

(2) 4.80×10^{-19} C

(3) 6.25 elementary charges

(4) 1.60 elementary charges

8.04 A: (2) all other choices require fractions of an elementary charge, while choice (2) is an integer multiple (3e) of elementary charges.

8.05 Q: What is the net electrical charge on a magnesium ion that is formed when a neutral magnesium atom loses two electrons?

(1) -3.2×10^{-19} C

(2) -1.6×10^{-19} C

(3) $+1.6 \times 10^{-19}$ C

(4) $+3.2 \times 10^{-19}$ C

8.05 A: (4) the net charge must be +2e, or $2(1.6 \times 10^{-19}$ C)$=3.2 \times 10^{-19}$ C.

Conductors and Insulators

Certain materials allow electric charges to move freely. These are called **conductors**. Examples of good conductors include metals such as gold, copper, silver, and aluminum. In contrast, materials in which electric charges cannot move freely are known as **insulators**. Good insulators include materials such as glass, plastic, and rubber. Metals are better conductors of electricity compared to insulators because metals contain more free electrons.

Conductors and insulators are characterized by their resistivity, or ability to resist movement of charge. Materials with high resistivities are good insulators. Materials with low resistivities are good conductors.

Semiconductors are materials which, in pure form, are good insulators. However, by adding small amounts of impurities known as dopants, their resistivities can be lowered significantly until they become good conductors.

Charging by Conduction

Materials can be charged by contact, or **conduction**. If you take a balloon and rub it against your hair, some of the electrons from the atoms in your hair are transferred to the balloon. The balloon now has extra electrons, and therefore has a net negative charge. Your hair has a deficiency of electrons, therefore it now has a net positive charge.

Much like momentum and energy, charge is also conserved. Continuing our hair and balloon example, the magnitude of the net positive charge on your hair is equal to the magnitude of the net negative charge on the balloon. The total charge of the hair/balloon system remains zero (neutral). For every extra electron (negative charge) on the balloon, there is a corresponding missing electron (positive charge) in your hair. This known as the law of conservation of charge.

Conductors can also be charged by contact. If a charged conductor is brought into conduct with an identical neutral conductor, the net charge will be shared across the two conductors.

8.06 Q: If a conductor carrying a net charge of 8 elementary charges is brought into contact with an identical conductor with no net charge, what will be the charge on each conductor after they are separated?

8.06 A: Each conductor will have a charge of 4 elementary charges.

8.07 Q: What is the net charge (in coulombs) on each conductor after they are separated?

8.07 A: $q = 4e = 4(1.6 \times 10^{-19} \text{ C}) = 6.4 \times 10^{-19} \text{ C}$

8.08 Q: Metal sphere A has a charge of −2 units and an identical metal sphere, B, has a charge of −4 units. If the spheres are brought into contact with each other and then separated, the charge on sphere B will be

(1) 0 units

(2) -2 units

(3) -3 units

(4) +4 units

8.08 A: (3) -3 units.

8.09 Q: Compared to insulators, metals are better conductors of electricity because metals contain more free

(1) protons

(2) electrons

(3) positive ions

(4) negative ions

8.09 A: (2) electrons.

A simple tool used to detect small electric charges known as an **electroscope** functions on the basis of conduction. The electroscope consists of a conducting rod attached to two thin conducting leaves at one end and isolated from surrounding charges by an insulating stopper placed in a flask. If a charged object is placed in contact with the conducting rod, part of the charge is transferred to the rod. Because the rod and leaves form a conducting path and like charges repel each other, the charges are distributed equally along the entire rod and leaf apparatus. The leaves, having like charges, repel each other, with larger charges providing greater leaf separation!

Charging by Induction

Conductors can also be charged without coming into contact with another charged object in a process known as charging by **induction**. This is accomplished by placing the conductor near a charged object and grounding the conductor. To understand charging by induction, you must first realize that when an object is connected to the earth by a conducting path, known as grounding, the earth acts like an infinite source for providing or accepting excess electrons.

To charge a conductor by induction, you first bring it close to another charged object. When the conductor is close to the charged object, any free electrons on the conductor will move toward the charged object if the object is positively charged (since opposite charges attract) or away from the charged object if the object is negatively charged (since like charges repel).

If the conductor is then "grounded" by means of a conducting path to the earth, the excess charge is compensated for by means of electron transfer to or from earth. Then the ground connection is severed. When the originally charged object is moved far away from the conductor, the charges in the conductor redistribute, leaving a net charge on the conductor as shown.

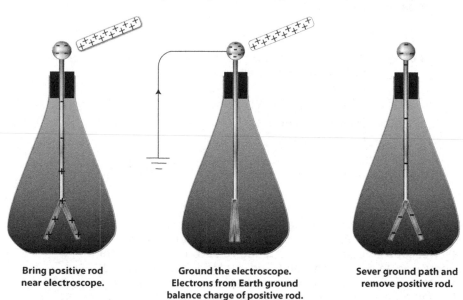

Bring positive rod near electroscope.

Ground the electroscope. Electrons from Earth ground balance charge of positive rod.

Sever ground path and remove positive rod.

You can also induce a charge in a charged region in a neutral object by bringing a strong positive or negative charge close to that object. In such cases, the electrons in the neutral object tend to move toward a strong positive charge, or away from a large negative charge. Though the object itself remains neutral, portions of the object are more positive or negative than other parts. In this way, you can attract a neutral object by bringing a charged object close to it, positive or negative. Put another way, a positively charged object can be attracted to both a negatively charged object and a neutral object, and a negatively charged object can be attracted to both a positively charged object and a neutral object.

For this reason, the only way to tell if an object is charged is by repulsion. A positively charge object can only be repelled by another positive charge and a negatively charged object can only be repelled by another negative charge.

8.10 Q: A positively charged glass rod attracts object X. The net charge of object X

(1) may be zero or negative

(2) may be zero or positive

(3) must be negative

(4) must be positive

8.10 A: (1) a positively charged rod can attract a neutral object or a negatively charged object.

8.11 Q: The diagram below shows three neutral metal spheres, x, y, and z, in contact and on insulating stands.

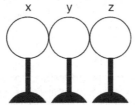

Which diagram best represents the charge distribution on the spheres when a positively charged rod is brought near sphere x, but does not touch it?

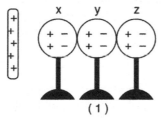

(1)

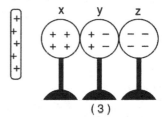

(3)

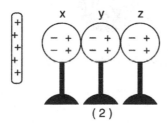

(2)

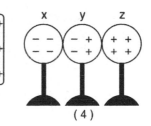

(4)

8.11 A: (4) is the correct answer.

Coulomb's Law

We know that like charges repel and opposite charges attract. In order for charges to repel or attract, they apply a force upon each either. Similar to the manner in which the force of attraction between two masses is determined by the amount of mass and the distance between the masses, as described by Newton's Law of Universal Gravitation, the force of attraction or repulsion is determined by the amount of charge and the distance between the charges.

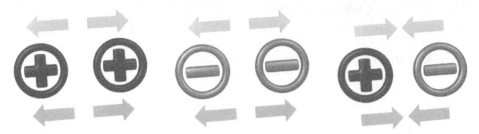

The magnitude of the electrostatic force is described by Coulomb's Law, which states that the magnitude of the electrostatic force (F_e) between two objects is equal to a constant, k, multiplied by each of the two charges, q_1 and q_2, and divided by the square of the distance between the charges (r^2). The constant k is known as the **electrostatic constant** and is given on the reference table as $k = 8.99 \times 10^9$ N·m²/C²

$$F_e = \frac{kq_1q_2}{r^2}$$

Notice how similar this formula is to the formula for the gravitational force! Both Newton's Law of Universal Gravitation and Coulomb's Law follow the inverse-square relationship, a pattern that repeats many times over in physics. The further you get from the charges, the weaker the electrostatic force. If you were to double the distance from a charge, you would quarter the electrostatic force on a charge.

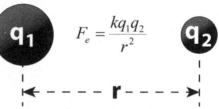

Formally, a positive value for the electrostatic force indicates that the force is a repelling force, while a negative value for the electrostatic force indicates that the force is an attractive force. Because force is a vector, you must assign a direction to it. To determine the direction of the force vector, once you have calculated its magnitude, use common sense to tell you the direction on each charged object. If the objects have opposite charges, they attract each other, and if they have like charges, they repel each other.

8.12 Q: Three protons are separated from a single electron by a distance of 1×10^{-6} m. Find the electrostatic force between them. Is this force attractive or repulsive?

8.12 A: $q_1 = 3 \text{ protons} = 3(1.6 \times 10^{-19} C) = 4.8 \times 10^{-19} C$

$q_2 = 1 \text{ electron} = 1(-1.6 \times 10^{-19} C) = -1.6 \times 10^{-19} C$

$$F_e = \frac{kq_1 q_2}{r^2} = \frac{(8.99 \times 10^9 \frac{N \cdot m^2}{C^2})(4.8 \times 10^{-19} C)(-1.6 \times 10^{-19} C)}{(1 \times 10^{-6} m)^2}$$

$$F_e = -6.9 \times 10^{-16} N$$

8.13 Q: A distance of 1.0 meter separates the centers of two small charged spheres. The spheres exert gravitational force F_g and electrostatic force F_e on each other. If the distance between the spheres' centers is increased to 3.0 meters, the gravitational force and electrostatic force, respectively, may be represented as

(1) $F_g/9$ and $F_e/9$

(2) $F_g/3$ and $F_e/3$

(3) $3F_g$ and $3F_e$

(4) $9F_g$ and $9F_e$

8.13 A: (1) due to the inverse square law relationships.

8.14 Q: A beam of electrons is directed into the electric field between two oppositely charged parallel plates, as shown in the diagram below.

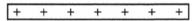

Electron beam

$\longrightarrow$

The electrostatic force exerted on the electrons by the electric field is directed

(1) into the page

(2) out of the page

(3) toward the bottom of the page

(4) toward the top of the page

8.14 A: (4) toward the top of the page because the electron beam is negative, and will be attracted by the positively charged upper plate and repelled by the negatively charged lower plate.

8.15 Q: The centers of two small charged particles are separated by a distance of 1.2×10^{-4} meter. The charges on the particles are $+8.0 \times 10^{-19}$ coulomb and $+4.8 \times 10^{-19}$ coulomb, respectively.

(A) Calculate the magnitude of the electrostatic force between these two particles.

(B) Sketch a graph showing the relationship between the magnitude of the electrostatic force between the two charged particles and the distance between the centers of the particles.

8.15 A: (A) $F_e = \dfrac{kq_1q_2}{r^2} = \dfrac{(8.99 \times 10^9\ \frac{N \bullet m^2}{C^2})(8.0 \times 10^{-19}\,C)(4.8 \times 10^{-19}\,C)}{(1.2 \times 10^{-4}\,m)^2}$

$F_e = 2.4 \times 10^{-19}\,N$

(B)

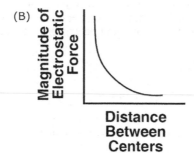

8.16 Q: The diagram below shows a beam of electrons fired through the region between two oppositely charged parallel plates in a cathode ray tube.

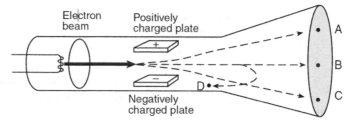

After passing between the charged plates, the electrons will most likely travel path

(1) A

(2) B

(3) C

(4) D

8.16 A: (1) A

Electric Fields

Similar to gravity, the electrostatic force is a non-contact force. Charged objects do not have to be in contact with each other to exert a force on each other. Somehow, a charged object feels the effect of another charged object through space. The property of space that allows a charged object to feel a force is a concept called the electric field. Although we cannot see an electric field, we can detect its presence by placing a positive test charge at various points in space and measuring the force the test charge feels.

While looking at gravity, the gravitational field strength was the amount of force observed by a mass per unit mass. In similar fashion, the electric field strength is the amount of electrostatic force observed by a charge per unit charge. Therefore, the electric field strength, E, is the electrostatic force observed at a given point in space divided by the test charge itself. Electric field strength is measured in Newtons per Coulomb (N/C).

$$E = \frac{F_e}{q}$$

8.17 Q: Two oppositely charged parallel metal plates, 1.00 centimeter apart, exert a force with a magnitude of 3.60 × 10⁻¹⁵ newtons on an electron placed between the plates. Calculate the magnitude of the electric field strength between the plates.

8.17 A: $E = \frac{F_e}{q} = \frac{3.6 \times 10^{-15}\, N}{1.6 \times 10^{-19}\, C} = 2.25 \times 10^4\, {}^N\!/\!_C$

8.18 Q: Which quantity and unit are correctly paired?
(1) resistivity and Ω/m
(2) potential difference and eV
(3) current and C•s
(4) electric field strength and N/C

8.18 A: (4) electric field strength and N/C.

8.19 Q: What is the magnitude of the electric field intensity at a point where a proton experiences an electrostatic force of magnitude 2.30×10⁻²⁵ newtons?
(1) 3.68×10⁻⁴⁴ N/C
(2) 1.44×10⁻⁶ N/C
(3) 3.68×10⁶ N/C
(4) 1.44×10⁴⁴ N/C

8.19 A: (2) $E = \dfrac{F_e}{q} = \dfrac{2.3 \times 10^{-25}\,N}{1.6 \times 10^{-19}\,C} = 1.44 \times 10^{-6}\,{}^N\!/_C$

8.20 Q: The diagram below represents an electron within an electric field between two parallel plates that are charged with a potential difference of 40.0 volts.

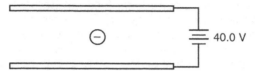

If the magnitude of the electric force on the electron is 2.00×10^{-15} newtons, the magnitude of the electric field strength between the charged plates is

(1) 3.20×10^{-34} N/C

(2) 2.00×10^{-14} N/C

(3) 1.25×10^{4} N/C

(4) 2.00×10^{16} N/C

8.20 A: (3) $E = \dfrac{F_e}{q} = \dfrac{2 \times 10^{-15}\,N}{1.6 \times 10^{-19}\,C} = 1.25 \times 10^{4}\,{}^N\!/_C$

Since we can't actually see the electric field, we can draw electric field lines to help us visualize the force a charge would feel if placed at a specific position in space. These lines show the direction of the electric force a positively charged particle would feel at that point. The more dense the lines are, the stronger the force a charged particle would feel, therefore the stronger the electric field. As the lines get further apart, the strength of the electric force a charged particle would feel is smaller, therefore the electric field is smaller.

By convention, electric field lines are drawn showing the direction of force on a positive charge. Therefore, to draw electric field lines for a system of charges, follow these basic rules:

1. Electric field lines point away from positive charges and toward negative charges.
2. Electric field lines never cross.
3. Electric field lines always intersect conductors at right angles to the surface.
4. Stronger fields have closer lines.
5. Field strength and line density decreases as you move away from the charges.

Let's take a look at a few examples of electric field lines, starting with isolated positive (left) and negative (right) charges. Notice that for each charge, the lines radiate outward or inward spherically. The lines point away from the positive charge, since a positive test charge placed in the field (near the fixed charge) would feel a repelling force. The lines point in toward the negative fixed charge, since a positive test charge would feel an attractive force.

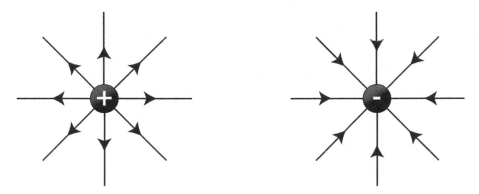

If you have both positive and negative charges in close proximity, you follow the same basic procedure:

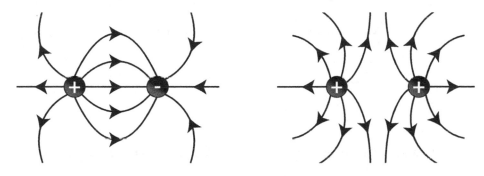

8.21 Q: Two small metallic spheres, A and B, are separated by a distance of 4.0×10^{-1} meter, as shown. The charge on each sphere is $+1.0 \times 10^{-6}$ coulomb. Point P is located near the spheres.

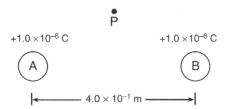

(A) What is the magnitude of the electrostatic force between the two charged spheres?

(1) 2.2×10^{-2} N

(2) 5.6×10^{-2} N

(3) 2.2×10^{4} N

(4) 5.6×10^{4} N

(B) Which arrow best represents the direction of the resultant electric field at point P due to the charges on spheres A and B?

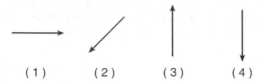

(1)　　　(2)　　　(3)　　　(4)

8.21 A:　(A) 2 $F_e = \dfrac{kq_1q_2}{r^2} = \dfrac{(8.99 \times 10^9 \; \frac{N \bullet m^2}{C^2})(1.0 \times 10^{-6}C)(1.0 \times 10^{-6}C)}{(4 \times 10^{-1}m)^2}$

$F_e = 0.056N$

(B) Correct answer is 3.

8.22 Q:　In the diagram below, P is a point near a negatively charged sphere.

Which vector best represents the direction of the electric field at point P?

(1)　　　(2)　　　(3)　　　(4)

8.22 A:　Correct answer is (1). Electric field lines point in toward negative charges.

8.23 Q:　Sketch at least four electric field lines with arrowheads that represent the electric field around a negatively charged conducting sphere.

8.23 A:

8.24 Q: The centers of two small charged particles are separated by a distance of 1.2×10^{-4} meter. The charges on the particles are $+8.0\times10^{-19}$ coulomb and $+4.8\times10^{-19}$ coulomb, respectively. Sketch at least four electric field lines in the region between the two positively charged particles.

8.24 A:

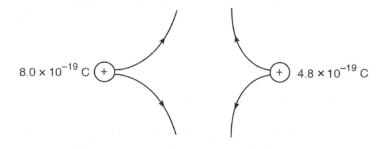

8.0×10^{-19} C 4.8×10^{-19} C

8.25 Q: Which graph best represents the relationship between the magnitude of the electric field strength, E, around a point charge and the distance, r, from the point charge?

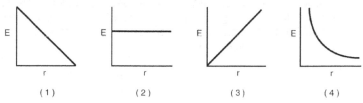

(1)   (2)   (3)   (4)

8.25 A: Correct answer is 4.

Because gravity and electrostatics have so many similarities, let's take a minute to do a quick comparison of electrostatics and gravity.

Electrostatics	Gravity
Force: $F_e = \dfrac{kq_1q_2}{r^2}$	**Force:** $F_g = \dfrac{Gm_1m_2}{r^2}$
Field Strength: $E = \dfrac{F_e}{q}$	**Field Strength:** $g = \dfrac{F_g}{m}$
Field Strength: $E = \dfrac{kq}{r^2}$	**Field Strength:** $g = \dfrac{Gm}{r^2}$
Constant: k=8.99×10^9 N·m²/C²	**Constant:** G=6.67×10^{-11} N·m²/kg²
Charge Units: Coulombs	**Mass Units:** kilograms

What is the big difference between electrostatics and gravity? The gravitational force can only attract, while the electrostatic force can both attract and repel. Notice again that both the electric field strength and the gravitational field strength follow the inverse-square law relationship. Field strength is inversely related to the square of the distance.

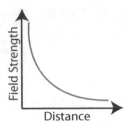

Electric Potential Difference

When we lifted an object against the force of gravity by applying a force over a distance, we did work to give that object gravitational potential energy. The same concept applies to electric fields as well. If you move a charge against an electric field, you must apply a force for some distance, therefore you do work and give it electrical potential energy. The work done per unit charge in moving a charge between two points in an electric field is known as the **electric potential difference**, (V). The units of electric potential are volts, where a volt is equal to 1 Joule per Coulomb. Therefore, if you do 1 Joule of work in moving a charge of 1 Coulomb in an electric field, the electric potential difference between those points would be 1 volt. This is given to you in the reference table as:

$$V = \frac{W}{q}$$

V in this formula is potential difference (in volts), W is work or electrical energy (in Joules), and q is your charge (in Coulombs). Let's take a look at some sample problems.

8.26 Q: A potential difference of 10 volts exists between two points, A and B, within an electric field. What is the magnitude of charge that requires 2.0×10^{-2} joules of work to move it from A to B?

8.26 A: $V = \frac{W}{q}$

$q = \frac{W}{V} = \frac{2 \times 10^{-2} J}{10V} = 2 \times 10^{-3} C$

8.27 Q: How much electrical energy is required to move a 4.00-microcoulomb charge through a potential difference of 36.0 volts?

(1) 9.00×10^6 J

(2) 144 J

(3) 1.44×10^{-4} J

(4) 1.11×10^{-7} J

8.27 A: (3) $V = \dfrac{W}{q}$

$$W = qV = (4 \times 10^{-6} C)(36V) = 1.44 \times 10^{-4} J$$

8.28 Q: If 1.0 joule of work is required to move 1.0 coulomb of charge between two points in an electric field, the potential difference between the two points is

(1) 1.0×10^0 V

(2) 9.0×10^9 V

(3) 6.3×10^{18} V

(4) 1.6×10^{-19} V

8.28 A: (1) $V = \dfrac{W}{q} = \dfrac{1J}{1C} = 1V = 1.0 \times 10^0 V$

8.29 Q: If 60 joules of work is required to move 5 coulombs of charge between two points in an electric field, what is the potential difference between these points?

(1) 5 V

(2) 12 V

(3) 60 V

(4) 300 V

8.29 A: (2) $V = \dfrac{W}{q} = \dfrac{60J}{5C} = 12V$

8.30 Q: In an electric field, 0.90 joules of work is required to bring 0.45 coulombs of charge from point A to point B. What is the electric potential difference between points A and B?

(1) 5.0 V

(2) 2.0 V

(3) 0.50 V

(4) 0.41 V

8.30 A: (2) $V = \dfrac{W}{q} = \dfrac{0.90J}{0.45C} = 2V$

When dealing with electrostatics, often times the amount of electric energy or work done on a charge is a very small portion of a Joule. Dealing with such small numbers is cumbersome, so physicists devised an alternate unit for electrical energy and work that can be more convenient than the Joule. This unit, known as the electronvolt (eV), is the amount of work done in moving an elementary charge through a potential difference of 1V. One electron-volt, therefore, is equivalent to one volt multiplied by one elementary charge (in Coulombs): 1 eV = 1.6×10^{-19} Joules.

8.31 Q: A proton is moved through a potential difference of 10 volts in an electric field. How much work, in electronvolts, was required to move this proton?

8.31 A: $V = \dfrac{W}{q}$

$W = qV = (1e)(10V) = 10eV$

Parallel Plates

If you know the potential difference between two parallel plates, you can easily calculate the electric field strength between the plates. As long as you're not near the edge of the plates, the electric field is constant between the plates and its strength is given by the equation:

$$E = \frac{V}{d}$$

You'll note that with the potential difference V in volts, and the distance between the plates in meters, units for the electric field strength are volts per meter [V/m]. Previously, we stated that the units for electric field strength were newtons per Coulomb [N/C]. It is easy to show these are equivalent:

$$\frac{N}{C} = \frac{N \bullet m}{C \bullet m} = \frac{J}{C \bullet m} = \frac{J/C}{m} = \frac{V}{m}$$

8.32 Q: The magnitude of the electric field strength between two oppositely charged parallel metal plates is 2.0×10^3 newtons per coulomb. Point P is located midway between the plates.

(A) Sketch at least five electric field lines to represent the field between the two oppositely charged plates.

(B) An electron is located at point P between the plates. Calculate the magnitude of the force exerted on the electron by the electric field.

8.32 A: (A)

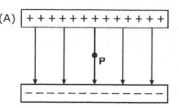

(B) $E = \dfrac{F_e}{q}$

$$F_e = qE = (1.6 \times 10^{-19}\, C)(2 \times 10^3 \, {}^N\!/\!_C) = 3.2 \times 10^{-16}\, N$$

8.33 Q: A moving electron is deflected by two oppositely charged parallel plates, as shown in the diagram below.

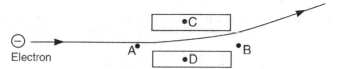

The electric field between the plates is directed from

(1) A to B
(2) B to A
(3) C to D
(4) D to C

8.33 A: (3) C to D because the electron feels a force opposite the direction of the electric field due to its negative charge.

8.34 Q: An electron is located in the electric field between two parallel metal plates as shown in the diagram below.

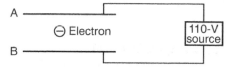

If the electron is attracted to plate A, then plate A is charged

(1) positively, and the electric field runs from plate A to plate B
(2) positively, and the electric field runs from plate B to plate A
(3) negatively, and the electric field runs from plate A to plate B
(4) negatively, and the electric field runs from plate B to plate A

8.34 A: Correct answer is 1.

8.35 Q: An electron placed between oppositely charged parallel plates moves toward plate A, as represented in the diagram below.

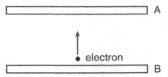

What is the direction of the electric field between the plates?

(1) toward plate A

(2) toward plate B

(3) into the page

(4) out of the page

8.35 A: (2) toward plate B.

8.36 Q: The diagram below represents two electrons, e_1 and e_2, located between two oppositely charged parallel plates.

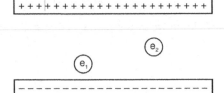

Compare the magnitude of the force exerted by the electric field on e_1 to the magnitude of the force exerted by the electric field on e_2.

8.36 A: The forces are the same because the electric field between two parallel plates is constant.

Equipotential Lines

Much like looking at a topographic map which shows you lines of equal altitude, or equal gravitational potential energy, we can make a map of the electric field and connect points of equal electrical potential. These lines, known as **equipotential lines**, always cross electrical field lines at right angles, and show positions in space with constant electrical potential. If you move a charged particle in space, and it always stays on an equipotential line, no work will be done.

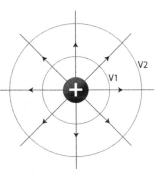

You can find more practice problems on the APlusPhysics website at: http://www.aplusphysics.com/regents.

Chapter 9: Current Electricity

"Electricity can be dangerous. My nephew tried to stick a penny into a plug. Whoever said a penny doesn't go far didn't see him shoot across that floor. I told him he was grounded."

— Tim Allen

Objectives

1. Define and calculate electric current.
2. Define and calculate resistance using Ohm's law.
3. Explain the factors and calculate the resistance of a conductor.
4. Identify the path and direction of current flow in a circuit.
5. Draw and interpret schematic diagrams of circuits.
6. Effectively use and analyze voltmeters and ammeters.
7. Solve series and parallel circuit problems using VIRP tables.
8. Calculate equivalent resistances for resistors in both series and parallel configurations.
9. Calculate power and energy used in electric circuits.

Electric Current

Electric current is the flow of charge, much like water currents are the flow of water molecules. Water molecules tend to flow from areas of high gravitational potential energy to low gravitational potential energy. Electric currents flow from high electric potential to low electric potential. The greater the difference between the high and low potential, the more current that flows!

In a majority of electric currents, the moving charges are negative electrons. However, due to historical reasons dating back to Ben Franklin, we say that conventional current flows in the direction positive charges would move. Although inconvenient, it's fairly easy to keep straight if you just remember that the actual moving charges, the electrons, flow in a direction opposite that of the electric current. With this in mind, we can state that positive current flows from high potential to low potential, even though the charge carriers (electrons) actually flow from low to high potential.

Electric current (I) is measured in amperes (A), or amps, and can be calculated by finding the total amount of charge (Δq), in coulombs, which passes a specific point in a given time (t). Electric current can therefore be calculated as:

$$I = \frac{\Delta q}{t}$$

9.01 Q: A charge of 30 Coulombs passes through a 24-ohm resistor in 6.0 seconds. What is the current through the resistor?

9.01 A: $I = \dfrac{\Delta q}{t} = \dfrac{30C}{6s} = 5A$

9.02 Q: Charge flowing at the rate of 2.50×10^{16} elementary charges per second is equivalent to a current of

        (1) 2.50×10^{13} A

        (2) 6.25×10^{5} A

        (3) 4.00×10^{-3} A

        (4) 2.50×10^{-3} A

9.02 A: $I = \dfrac{\Delta q}{t} = \dfrac{(2.50 \times 10^{16})(1.6 \times 10^{-19}C)}{1s} = 4 \times 10^{-3}A$

9.03 Q: The current through a lightbulb is 2.0 amperes. How many coulombs of electric charge pass through the lightbulb in one minute?

        (1) 60 C

        (2) 2.0 C

(3) 120 C

(4) 240 C

9.03 A: (3) $I = \dfrac{\Delta q}{t}$

$$\Delta q = It = (2A)(60s) = 120C$$

9.04 Q: A 1.5-volt, AAA cell supplies 750 milliamperes of current through a flashlight bulb for 5 minutes, while a 1.5-volt, C cell supplies 750 milliamperes of current through the same flashlight bulb for 20 minutes. Compared to the total charge transferred by the AAA cell through the bulb, the total charge transferred by the C cell through the bulb is

(1) half as great

(2) twice as great

(3) the same

(4) four times as great

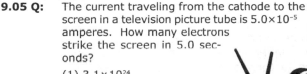

9.04 A: (4) If Δq=It, and both cells supply 0.750A but the C cell supplies the same current for four times as long, it must supply four times the total charge compared to the AAA cell.

9.05 Q: The current traveling from the cathode to the screen in a television picture tube is 5.0×10^{-5} amperes. How many electrons strike the screen in 5.0 seconds?

(1) 3.1×10^{24}

(2) 6.3×10^{18}

(3) 1.6×10^{15}

(4) 1.0×10^{5}

9.05 A: (3)

$$I = \dfrac{\Delta q}{t}$$

$$\Delta q = It = (5 \times 10^{-5}A)(5s) = 2.5 \times 10^{-4}C$$

$$2.5 \times 10^{-4}C \bullet \dfrac{1 \text{ electron}}{1.6 \times 10^{-19}C} = 1.6 \times 10^{15} \text{ electrons}$$

Resistance

Electrical charges can move easily in some materials (conductors) and less freely in others (insulators), as we learned previously. We describe a material's ability to conduct electric charge as **conductivity**. Good conductors have high conductivities. The conductivity of a material depends on:

1. Density of free charges available to move
2. Mobility of those free charges

In similar fashion, we describe a material's ability to resist the movement of electric charge using **resistivity**, symbolized with the Greek letter rho (ρ). Resistivity is measured in ohm-meters, which are represented by the Greek letter omega multiplied by meters (Ω•m). Both conductivity and resistivity are properties of a material.

When an object is created out of a material, the material's tendency to conduct electricity, or conductance, depends on the material's conductivity as well as the material's shape. For example, a hollow cylindrical pipe has a higher conductivity of water than a cylindrical pipe filled with cotton. However, the shape of the pipe also plays a role. A very thick but short pipe can conduct lots of water, yet a very narrow, very long pipe can't conduct as much water. Both geometry of the object and the object's composition influence its conductance.

Focusing on an object's ability to resist the flow of electrical charge, we find that objects made of high resistivity materials tend to impede electrical current flow and have a high resistance. Further, materials shaped into long, thin objects also increase an object's electrical resistance. Finally, objects typically exhibit higher resistivities at higher temperatures. We take all of these factors together to describe an object's resistance to the flow of electrical charge. Resistance is a functional property of an object that describes the object's ability to impede the flow of charge through it. Units of resistance are ohms (Ω).

For any given temperature, we can calculate an object's electrical resistance, in ohms, using the following formula, which can be found on your reference table.

$$R = \frac{\rho L}{A}$$

Resistivities at 20°C	
Material	**Resistivity (Ω•m)**
Aluminum	2.82×10^{-8}
Copper	1.72×10^{-8}
Gold	2.44×10^{-8}
Nichrome	$150. \times 10^{-8}$
Silver	1.59×10^{-8}
Tungsten	5.60×10^{-8}

In this formula, R is the resistance of the object, in ohms (Ω), rho (ρ) is the resistivity of the material the object is made out of, in ohm•meters (Ω•m), L is the length of the object, in meters, and A is the cross-sectional area of the object, in meters squared. Note that a table of material resistivities for a constant temperature is given to you on the reference table!

9.06 Q: A 3.50-meter length of wire with a cross-sectional area of 3.14×10^{-6} m^2 at 20° Celsius has a resistance of 0.0625 Ω. Determine the resistivity of the wire and the material it is made out of.

9.06 A:
$$R = \frac{\rho L}{A}$$

$$\rho = \frac{RA}{L} = \frac{(.0625\Omega)(3.14\times10^{-6}\,m^2)}{3.5m} = 5.6\times10^{-8}\,\Omega\bullet m$$

Material must be tungsten.

9.07 Q: The electrical resistance of a metallic conductor is inversely proportional to its

(1) temperature

(2) length

(3) cross-sectional area

(4) resistivity

9.07 A: (3) straight from the formula.

9.08 Q: At 20°C, four conducting wires made of different materials have the same length and the same diameter. Which wire has the least resistance?

(1) aluminum

(2) gold

(3) nichrome

(4) tungsten

9.08 A: (2) gold because it has the lowest resistivity.

9.09 Q: A length of copper wire and a 1.00-meter-long silver wire have the same cross-sectional area and resistance at 20°C. Calculate the length of the copper wire.

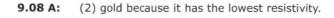

9.09 A: $R = \left(\dfrac{\rho L}{A}\right)_{copper} = \left(\dfrac{\rho L}{A}\right)_{silver}$

$$R = \dfrac{\rho_{copper} L_{copper}}{A} = \dfrac{\rho_{silver} L_{silver}}{A}$$

$$L_{copper} = \dfrac{\rho_{silver} L_{silver}}{\rho_{copper}} = \dfrac{(1.59 \times 10^{-8}\,\Omega m)(1m)}{1.72 \times 10^{-8}\,\Omega m}$$

$$L_{copper} = 0.924m$$

9.10 Q: A 10-meter length of copper wire is at 20°C. The radius of the wire is 1.0×10⁻³ meter.

Cross Section of Copper Wire

r = 1.0 × 10⁻³ m

(A) Determine the cross-sectional area of the wire.

(B) Calculate the resistance of the wire.

9.10 A: (A) $Area_{circle} = \pi r^2 = \pi(1.0 \times 10^{-3}\,m)^2 = 3.14 \times 10^{-6}\,m^2$

(B) $R = \dfrac{\rho L}{A} = \dfrac{(1.72 \times 10^{-8}\,\Omega m)(10m)}{3.14 \times 10^{-6}\,m^2} = 5.5 \times 10^{-2}\,\Omega$

Ohm's Law

If resistance opposes current flow, and potential difference promotes current flow, it only makes sense that these quantities must somehow be related. George Ohm studied and quantified these relationships for conductors and resistors in a famous formula now known as Ohm's Law:

$$R = \frac{V}{I}$$

Ohm's Law may make more qualitative sense if we re-arrange it slightly:

$$V = IR$$

Now it's easy to see that the current flowing through a conductor or resistor (in amps) is equal to the potential difference across the object (in volts) divided by the resistance of the object (in ohms). If you want a large current to flow, you require a large potential difference (such as a large battery), and/or a very small resistance.

9.11 Q: The current in a wire is 24 amperes when connected to a 1.5 volt battery. Find the resistance of the wire.

9.11 A: $R = \dfrac{V}{I} = \dfrac{1.5V}{24A} = 0.0625\Omega$

9.12 Q: In a simple electric circuit, a 24-ohm resistor is connected across a 6-volt battery. What is the current in the circuit?

(1) 1.0 A

(2) 0.25 A

(3) 140 A

(4) 4.0 A

9.12 A: (2) $I = \dfrac{V}{R} = \dfrac{6V}{24\Omega} = 0.25A$

9.13 Q: A constant potential difference is applied across a variable resistor held at constant temperature. Which graph best represents the relationship between the resistance of the variable resistor and the current through it?

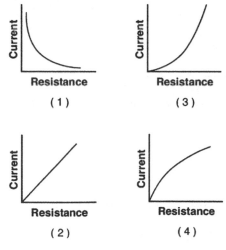

9.13 A: (1) due to Ohm's Law (I=V/R).

9.14 Q: What is the current in a 100-ohm resistor connected to a 0.40-volt source of potential difference?

(1) 250 mA

(2) 40 mA

(3) 2.5 mA

(4) 4.0 mA

9.14 A: (4) $I = \dfrac{V}{R} = \dfrac{0.40V}{100\Omega} = 0.004\,A = 4mA$

Note: Ohm's Law isn't truly a law of physics -- not all materials obey this relationship. It is, however, a very useful empirical relationship that accurately describes key electrical characteristics of conductors and resistors. One way to test if a material is ohmic (if it follows Ohm's Law) is to graph the voltage vs. current flow through the material. If the material obeys Ohm's Law, you get a linear relationship, where the slope of the line is equal to the material's resistance.

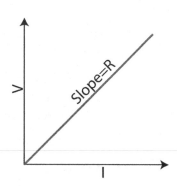

9.15 Q: The graph below represents the relationship between the potential difference (V) across a resistor and the current (I) through the resistor.

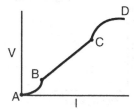

Through which entire interval does the resistor obey Ohm's law?

(1) AB

(2) BC

(3) CD

(4) AD

9.15 A: (2) BC because the graph is linear in this interval.

Electrical Circuits

An **electrical circuit** is a closed loop path through which current can flow. An electrical circuit can be made up of almost any materials (including humans if we're not careful), but practically speaking, they are typically comprised of electrical devices such as wires, batteries, resistors, and switches. Conventional current will flow through a complete closed-loop (closed circuit) from high potential to low potential. Therefore, electrons actually flow in the opposite direction, from low potential to high potential. If the path isn't a closed loop (and is, instead, an open circuit), no current will flow.

Electric circuits, which are three-dimensional constructs, are typically represented in two dimensions using diagrams known as **circuit schematics**.

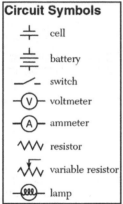

Circuit Symbols	
$\perp$	cell
$\equiv$	battery
$\diagup$	switch
$-(V)-$	voltmeter
$-(A)-$	ammeter
$\diagdown\!\!\diagdown\!\!\diagup$	resistor
$\diagdown\!\!\!\diagup\!\!\!\diagdown$	variable resistor
$-\!\!\bigcirc\!\!-$	lamp

These schematics are simplified, standardized representations in which common circuit elements are represented with specific symbols, and wires connecting the elements in the circuit are represented by lines. Basic circuit schematic symbols are shown in the Physics Reference Table.

In order for current to flow through a circuit, you must have a source of potential difference. Typical sources of potential difference are voltaic cells, batteries (which are just two or more cells connected together), and power (voltage) supplies. We often times refer to voltaic cells as batteries in common terminology. In drawing a cell or battery on a circuit schematic, remember that the longer side of the symbol is the positive terminal.

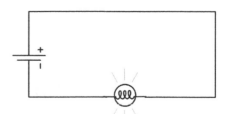

AA Voltaic Cell   9V Battery

Electric circuits must form a complete conducting path in order for current to flow. In the example circuit shown below left, the circuit is incomplete because the switch is open, therefore no current will flow and the lamp will not light. In the circuit below right, however, the switch is closed, creating a closed loop path. Current will flow and the lamp will light up.

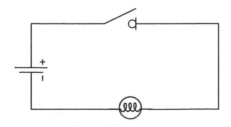

Note that in the picture above right, conventional current will flow from positive to negative, creating a clockwise current path in the circuit. The actual electrons in the wire, however, are flowing in the opposite direction, or counter-clockwise.

Energy & Power

Just like mechanical power is the rate at which mechanical energy is expended, **electrical power** is the rate at which electrical energy is expended. We learned previously that when you do work on something you change its energy and that electrical work or energy is equal to charge times potential difference. Therefore, we can write the equation for electrical power as:

$$P = \frac{W}{t} = \frac{qV}{t}$$

We also know, however, that the amount of charge moving past a point per given unit of time is current, therefore we can continue the derivation as follows:

$$P = \left(\frac{q}{t}\right)V = IV$$

So electrical power expended in a circuit is the electrical current multiplied by potential difference (voltage). Using Ohm's Law, we can expand this even further to provide several different methods for calculating electrical power dissipated by a resistor:

$$P = VI = I^2R = \frac{V^2}{R}$$

Of course, conservation of energy still applies, so the energy used in the resistor is converted into heat (in most cases) and light, or it can be used to do work. Let's see if we can't put this knowledge to use in a practical application.

9.16 Q: A 110-volt toaster oven draws a current of 6 amps on its highest setting as it converts electrical energy into thermal energy. What is the toaster's maximum power rating?

9.16 A: $P = VI = (110V)(6A) = 660W$

9.17 Q: An electric iron operating at 120 volts draws 10 amperes of current. How much heat energy is delivered by the iron in 30 seconds?

(1) 3.0×10^2 J

(2) 1.2×10^3 J

(3) 3.6×10^3 J

(4) 3.6×10^4 J

9.17 A: (4) $W = Pt = VIt = (120V)(10A)(30s) = 3.6 \times 10^4 J$

9.18 Q: One watt is equivalent to one

(1) N·m

(2) N/m

(3) J·s

(4) J/s

9.18 A: (4) J/s, since Power is W/t, and the unit of work is the joule, and the unit of time is the second.

9.19 Q: A potential drop of 50 volts is measured across a 250-ohm resistor. What is the power developed in the resistor?

(1) 0.20 W

(2) 5.0 W

(3) 10 W

(4) 50 W

9.19 A: $P = \dfrac{V^2}{R} = \dfrac{(50V)^2}{250\Omega} = 10W$

9.20 Q: What is the minimum information needed to determine the power dissipated in a resistor of unknown value?

(1) potential difference across the resistor, only

(2) current through the resistor, only

(3) current and potential difference, only

(4) current, potential difference, and time of operation

9.20 A: (3) current and potential difference, only (P=VI).

Voltmeters

Voltmeters are tools used to measure the potential difference between two points in a circuit. The voltmeter is connected in parallel with the element to be measured, meaning an alternate current path around the element to be measured and through the voltmeter is created. You have connected a voltmeter correctly if you can remove the voltmeter from the circuit without breaking the circuit. In the diagram at right, a voltmeter is connected to correctly measure the potential difference across the lamp. Voltmeters have very high resistance so as to minimize the current flow through the voltmeter and the voltmeter's impact on the circuit.

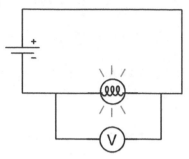

Ammeters

Ammeters are tools used to measure the current in a circuit. The ammeter is connected in series with the circuit, so that the current to be measured flows directly through the ammeter. The circuit must be broken to correctly insert an ammeter. Ammeters have very low resistance to minimize the potential drop through the ammeter and the ammeter's impact on the circuit, so inserting an ammeter into a circuit in parallel can result in extremely high currents and may destroy the ammeter. In the diagram at right, an ammeter is connected correctly to measure the current flowing through the circuit.

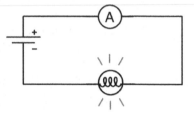

9.21 Q: In the electric circuit diagram, possible locations of an ammeter and voltmeter are indicated by circles 1, 2, 3, and 4. Where should an ammeter be located to correctly measure the total current and where should a voltmeter be located to correctly measure the total voltage?

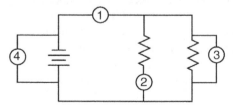

(1) ammeter at 1 and voltmeter at 4

(2) ammeter at 2 and voltmeter at 3

(3) ammeter at 3 and voltmeter at 4

(4) ammeter at 1 and voltmeter at 2

9.21 A: (1) To measure the total current, the ammeter must be placed at position 1, as all the current in the circuit must pass through this wire, and ammeters are always connected in series. To measure the total voltage in the circuit, the voltmeter could be placed at either position 3 or position 4. Voltmeters are always placed in parallel with the circuit element being analyzed, and positions 3 and 4 are equivalent because they are connected with wires (and potential is always the same anywhere in an ideal wire).

9.22 Q: Which circuit diagram below correctly shows the connection of ammeter A and voltmeter V to measure the current through and potential difference across resistor R?

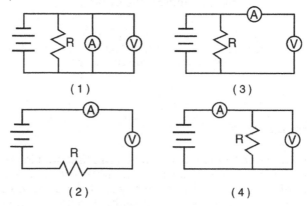

9.22 A: (4) shows an ammeter in series and a voltmeter in parallel with the resistor.

9.23 Q: A student uses a voltmeter to measure the potential difference across a resistor. To obtain a correct reading, the student must connect the voltmeter

(A) in parallel with the resistor

(B) in series with the resistor

(C) before connecting the other circuit components

(D) after connecting the other circuit components

9.23 A: (A) in parallel with the resistor.

9.24 Q: Which statement about ammeters and voltmeters is correct?

(1) The internal resistance of both meters should be low.

(2) Both meters should have a negligible effect on the circuit being measured.

(3) The potential drop across both meters should be made as large as possible.

(4) The scale range on both meters must be the same.

9.24 A: (2) Both meters should have a negligible effect on the circuit being measured.

9.25 Q: Compared to the resistance of the circuit being measured, the internal resistance of a voltmeter is designed to be very high so that the meter will draw

(1) no current from the circuit

(2) little current from the circuit

(3) most of the current from the circuit

(4) all the current from the circuit

9.25 A: (2) the voltmeter should draw as little current as possible from the circuit to minimize its effect on the circuit, but it does require some small amount of current to operate.

Series Circuits

Developing an understanding of circuits is the first step in learning about the modern-day electronic devices that dominate what is becoming known as the "Information Age." A basic circuit type, the **series circuit**, is a circuit in which there is only a single current path. Kirchhoff's Laws provide us the tools in order to analyze any type of circuit.

Kirchhoff's Current Law (KCL), named after German physicist Gustav Kirchhoff, states that the sum of all current entering any point in a circuit has to equal the sum of all current leaving any point in a circuit. More simply, this is another way of looking at the law of conservation of charge.

Kirchhoff's Voltage Law (KVL) states that the sum of all the potential drops in any closed loop of a circuit has to equal zero. More simply, KVL is a method of applying the law of conservation of energy to a circuit.

9.26 Q: A 3.0-ohm resistor and a 6.0-ohm resistor are connected in series in an operating electric circuit. If the current through the 3.0-ohm resistor is 4.0 amperes, what is the potential difference across the 6.0-ohm resistor?

9.26 A: First, let's draw a picture of the situation. If 4 amps of current is flowing through the 3-ohm resistor, then 4 amps of current must be flowing through the 6-ohm resistor according to Kirchhoff's Current Law. If we know the current and the resistance, we can calculate the voltage drop across the 6-ohm resistor using Ohm's Law: V=IR=(4A)(6Ω)=24V.

9.27 Q: The diagram below represents currents in a segment of an electric circuit.

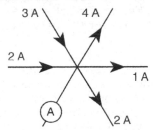

What is the reading of ammeter A?

(1) 1 A

(2) 2 A

(3) 3 A

(4) 4 A

9.27 A: (2) Since five amps plus the unknown current are coming in to the junction, and seven amps are leaving, KCL tells us the total current in must equal the total current out, therefore the unknown current must be two amps in to the junction.

Let's take a look at a sample circuit, consisting of three 2000-ohm (2Kohm) resistors:

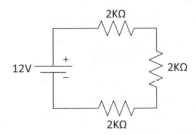

There is only a single current path in the circuit, which travels through all three resistors. Instead of using three separate 2KΩ (2000Ω) resistors, we could replace the three resistors with one single resistor having an equivalent resistance. To find the equivalent resistance of any number of series resistors, just add up their individual resistances:

$$R_{eq} = R_1 + R_2 + R_3 + ...$$
$$R_{eq} = 2000Ω + 2000Ω + 2000Ω$$
$$R_{eq} = 6000Ω = 6KΩ$$

Note that because there is only a single current path, the same current must flow through each of the resistors.

A simple and straightforward method for analyzing circuits involves creating a VIRP table for each circuit you encounter. Combining your knowledge of Ohm's Law, Kirchhoff's Current Law, Kirchhoff's Voltage Law, and equivalent resistance, you can use this table to solve for the details of any circuit.

A VIRP table describes the potential drop (V-voltage), current flow (I-current), resistance (R) and power dissipated (P-power) for each element in your circuit, as well as for the circuit as a whole. Let's use our circuit with the three 2000-ohm resistors as an example to demonstrate how a VIRP table is used. To create the VIRP table, first list the circuit elements, and total, in the rows of the table, then make columns for V, I, R, and P:

VIRP Table

	V	I	R	P
R_1				
R_2				
R_3				
Total				

Next, fill in the information in the table that is known. For example, we know the total voltage in the circuit (12V) provided by the battery, and we know the values for resistance for each of the individual resistors:

	V	I	R	P
R_1			2000Ω	
R_2			2000Ω	
R_3			2000Ω	
Total	12V			

Once the initial information has been filled in, you can also calculate the total resistance, or equivalent resistance, of the entire circuit. In this case, the equivalent resistance is 6000 ohms.

	V	I	R	P
R_1			2000Ω	
R_2			2000Ω	
R_3			2000Ω	
Total	12V		6000Ω	

Looking at the bottom (total) row of the table, both the voltage drop (V) and the resistance (R) are known. Using these two items, the total current flow in the circuit can be calculated using Ohm's Law.

$$I = \frac{V}{R} = \frac{12V}{6000\Omega} = 0.002\,A$$

The total power dissipated can also be calculated using any of the formulas for electrical power.

$$P = \frac{V^2}{R} = \frac{(12V)^2}{6000\Omega} = 0.024W$$

More information can now be completed in the VIRP table:

	V	I	R	P
R_1			2000Ω	
R_2			2000Ω	
R_3			2000Ω	
Total	12V	0.002A	6000Ω	0.024W

Because this is a series circuit, the total current has to be the same as the current through each individual element, so you can fill in the current through each of the individual resistors:

	V	I	R	P
R_1		0.002A	2000Ω	
R_2		0.002A	2000Ω	
R_3		0.002A	2000Ω	
Total	12V	0.002A	6000Ω	0.024W

Finally, for each element in the circuit, you now know the current flow and the resistance. Using this information, Ohm's Law can be applied to obtain the voltage drop (V=IR) across each resistor. Power can also be found for each element using $P=I^2R$ to complete the table.

	V	I	R	P
R_1	4V	0.002A	2000Ω	0.008W
R_2	4V	0.002A	2000Ω	0.008W
R_3	4V	0.002A	2000Ω	0.008W
Total	12V	0.002A	6000Ω	0.024W

So what does this table really tell us now that it's completely filled out? We know the potential drop across each resistor (4V), the current through each resistor (2 mA), and the power dissipated by each resistor (8 mW). In addition, we know the total potential drop for the entire circuit is 12V, and the entire circuit dissipated 24 mW of power. Note that for a series circuit, the sum of the individual voltage drops across each element equal the total potential difference in the circuit, the current is the same throughout the circuit, and the resistances and power dissipated values add up to the total resistance and total power dissipated. These are summarized for you on your reference table as follows:

$$I = I_1 = I_2 = I_3 = \dots$$
$$V = V_1 + V_2 + V_3 + \dots$$
$$R_{eq} = R_1 + R_2 + R_3 + \dots$$

9.28 Q: A 2.0-ohm resistor and a 4.0-ohm resistor are connected in series with a 12-volt battery. If the current through the 2.0-ohm resistor is 2.0 amperes, the current through the 4.0-ohm resistor is

(1) 1.0 A

(2) 2.0 A

(3) 3.0 A

(4) 4.0 A

9.28 A: (2) The current through a series circuit is the same everywhere, therefore the correct answer must be 2.0 amperes.

9.29 Q: In the circuit diagram below, two 4.0-ohm resistors are connected to a 16-volt battery as shown.

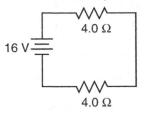

The rate at which electrical energy is expended in this circuit is

(1) 8.0 W

(2) 16 W

(3) 32 W

(4) 64 W

9.29 A: (3) 32W. Rate at which energy is expended is known as power.

	V	I	R	P
R₁	8V	2A	4Ω	16W
R₂	8V	2A	4Ω	16W
Total	16V	2A	8Ω	32W

9.30 Q: A 50-ohm resistor, an unknown resistor R, a 120-volt source, and an ammeter are connected in a complete circuit. The ammeter reads 0.50 ampere.

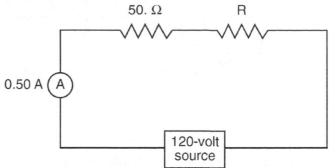

(A) Calculate the equivalent resistance of the circuit.

(B) Determine the resistance of resistor R.

(C) Calculate the power dissipated by the 50-ohm resistor.

9.30 A: (A) R_{eq} = 240Ω (B) R= 190Ω (C) $P_{50\Omega\ resistor}$= 12.5W

	V	I	R	P
R$_1$	25V	0.50A	50Ω	12.5W
R$_2$	95V	0.50A	190Ω	47.5W
Total	120V	0.50A	240Ω	60W

9.31 Q: What must be inserted between points A and B to establish a steady electric current in the incomplete circuit represented in the diagram?

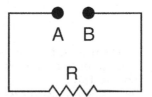

(1) switch

(2) voltmeter

(3) magnetic field source

(4) source of potential difference

9.31 A: (4) a source of potential difference is required to drive current.

9.32 Q: In the circuit represented by the diagram, what is the reading of voltmeter V?

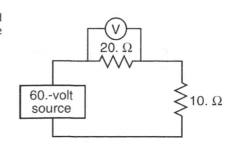

(1) 20 V

(2) 2.0 V

(3) 30 V

(4) 40 V

9.32 A: (4) Voltmeter reads potential difference across R$_1$ which is 40 V.

	V	I	R	P
R$_1$	40V	2A	20Ω	80W
R$_2$	20V	2A	10Ω	40W
Total	60V	2A	30Ω	120W

Parallel Circuits

Another basic circuit type is the **parallel circuit**, in which there is more than one current path. To analyze resistors in a series circuit, we found an equivalent resistance. We'll follow the same strategy in analyzing resistors in parallel.

Let's look at a circuit made of the same components we used in our exploration of series circuits, but now we'll connect our components so as to provide multiple current paths, creating a parallel circuit.

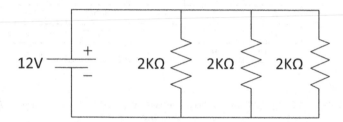

Notice that in this circuit, electricity can follow one of three different paths through each of the resistors. In many ways, this is similar to a river branching into three different smaller rivers. Each resistor, then, causes a potential drop (analogous to a waterfall), then the three rivers recombine before heading back to the battery, which we can think of like a pump, raising the river to a higher potential before sending it back on its looping path.

We can find the equivalent resistance of resistors in parallel using the formula:

$$\frac{1}{R_{eq}} = \frac{1}{R_1} + \frac{1}{R_2} + \frac{1}{R_3} + ...$$

Take care in using this equation, as it's easy to make errors in performing your calculations. Let's see if we can find the equivalent resistance for our sample circuit.

$$\frac{1}{R_{eq}} = \frac{1}{R_1} + \frac{1}{R_2} + \frac{1}{R_3} + \ldots$$

$$\frac{1}{R_{eq}} = \frac{1}{2000\Omega} + \frac{1}{2000\Omega} + \frac{1}{2000\Omega}$$

$$\frac{1}{R_{eq}} = 0.0015\,{}^{1}\!/_{\Omega}$$

$$R_{eq} = \frac{1}{0.0015\,{}^{1}\!/_{\Omega}} = 667\Omega$$

A VIRP table can again be used to analyze the circuit, beginning by filling in what is known directly from the circuit diagram.

VIRP Table

	V	**I**	**R**	**P**
R₁			**2000Ω**	
R₂			**2000Ω**	
R₃			**2000Ω**	
Total	**12V**			

You can also see from the circuit diagram that the potential drop across each resistor must be 12V, since the ends of each resistor are held at a 12-volt difference by the battery

	V	**I**	**R**	**P**
R₁	**12V**		2000Ω	
R₂	**12V**		2000Ω	
R₃	**12V**		2000Ω	
Total	12V			

Next, you can use Ohm's Law to fill in the current through each of the individual resistors since you know the voltage drop across each resistor (I=V/R) to find I=0.006A.

	V	**I**	**R**	**P**
R₁	12V	**0.006A**	2000Ω	
R₂	12V	**0.006A**	2000Ω	
R₃	12V	**0.006A**	2000Ω	
Total	12V			

Using Kirchhoff's Current Law, we can see that if 0.006A flows through each of the resistors, these currents all come together to form a total current of 0.018A.

	V	I	R	P
R₁	12V	0.006A	2000Ω	
R₂	12V	0.006A	2000Ω	
R₃	12V	0.006A	2000Ω	
Total	12V	**0.018A**		

Because each of the three resistors has the same resistance, it only makes sense that the current would be split evenly between them. You can confirm the earlier calculation of equivalent resistance by calculating the total resistance of the circuit using Ohm's Law: R=V/I=(12V/0.018A)=667Ω.

	V	I	R	P
R₁	12V	0.006A	2000Ω	
R₂	12V	0.006A	2000Ω	
R₃	12V	0.006A	2000Ω	
Total	12V	0.018A	**667Ω**	

Finally, you can complete the VIRP table using any of the three applicable equations for power dissipation to find:

	V	I	R	P
R₁	12V	0.006A	2000Ω	**0.072W**
R₂	12V	0.006A	2000Ω	**0.072W**
R₃	12V	0.006A	2000Ω	**0.072W**
Total	12V	0.018A	667Ω	**0.216W**

Note that for resistors in parallel, the equivalent resistance is always less than the resistance of any of the individual resistors. The potential difference across each of the resistors in parallel is the same, and the current through each of the resistors adds up to the total current. This is summarized for you on the reference table:

$$I = I_1 + I_2 + I_3 + ...$$
$$V = V_1 = V_2 = V_3 = ...$$
$$\frac{1}{R_{eq}} = \frac{1}{R_1} + \frac{1}{R_2} + \frac{1}{R_3} + ...$$

9.33 Q: A 15-ohm resistor, R_1, and a 30-ohm resistor, R_2, are to be connected in parallel between points A and B in a circuit containing a 90-volt battery.

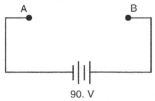

90. V

(A) Complete the diagram to show the two resistors connected in parallel between points A and B.

(B) Determine the potential difference across resistor R_1.

(C) Calculate the current in resistor R_1.

9.33 A: (A)

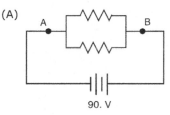

90. V

(B) Potential difference across R_1 is 90V.

(C) Current through resistor R_1 is 6A.

	V	I	R	P
R_1	90V	6A	15Ω	540W
R_2	90V	3A	30Ω	270W
Total	90V	9A	10Ω	810W

9.34 Q: Draw a diagram of an operating circuit that includes: a battery as a source of potential difference, two resistors in parallel with each other, and an ammeter that reads the total current in the circuit.

9.34 A:

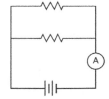

9.35 Q: Three identical lamps are connected in parallel with each other. If the resistance of each lamp is X ohms, what is the equivalent resistance of this parallel combination?

(1) X Ω

(2) X/3 Ω

(3) 3X Ω

(4) 3/X Ω

9.35 A: (2) X/3 Ω

$$\frac{1}{R_{eq}} = \frac{1}{R_1} + \frac{1}{R_2} + \frac{1}{R_3} + ...$$

$$\frac{1}{R_{eq}} = \frac{1}{X} + \frac{1}{X} + \frac{1}{X}$$

$$\frac{1}{R_{eq}} = \frac{3}{X}$$

$$R_{eq} = \frac{X}{3}$$

9.36 Q: Three resistors, 4 ohms, 6 ohms, and 8 ohms, are connected in parallel in an electric circuit. The equivalent resistance of the circuit is

(1) less than 4 Ω

(2) between 4 Ω and 8 Ω

(3) between 10 Ω and 18 Ω

(4) 18 Ω

9.36 A: (1) the equivalent resistance of resistors in parallel is always less than the value of the smallest resistor.

9.37 Q: A 3-ohm resistor, an unknown resistor, R, and two ammeters, A₁ and A₂, are connected as shown with a 12-volt source. Ammeter A₂ reads a current of 5 amperes.

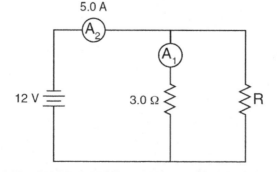

(A) Determine the equivalent resistance of the circuit.

(B) Calculate the current measured by ammeter A_1.

(C) Calculate the resistance of the unknown resistor, R.

9.37 A: (A) 2.4Ω (B) 4A (C) 12Ω

	V	I	R	P
R_1	12V	4A	3Ω	48W
R_2	12V	1A	12Ω	12W
Total	12V	5A	2.4Ω	60W

9.38 Q: The diagram below represents an electric circuit consisting of four resistors and a 12-volt battery.

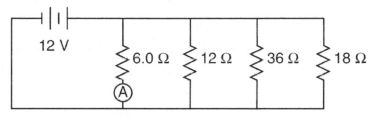

(A) What is the current measured by ammeter A?

(B) What is the equivalent resistance of this circuit?

(C) How much power is dissipated in the 36-ohm resistor?

9.38 A: (A) 2A (B) 3Ω (C) 4W

	V	I	R	P
R_1	12V	2A	6Ω	24W
R_2	12V	1A	12Ω	12W
R_3	12V	0.33A	36Ω	4W
R_4	12V	0.67A	18Ω	8W
Total	12V	4A	3Ω	48W

9.39 Q: A 20-ohm resistor and a 30-ohm resistor are connected in parallel to a 12-volt battery as shown. An ammeter is connected as shown.

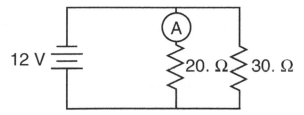

(A) What is the equivalent resistance of the circuit?
(B) What is the current reading of the ammeter?
(C) What is the power of the 30-ohm resistor?

9.39 A: (A) 12Ω   (B) 0.6A   (C) 4.8W

	V	I	R	P
R₁	12V	0.6A	20Ω	7.2W
R₂	12V	0.4A	30Ω	4.8W
Total	12V	1A	12Ω	12W

9.40 Q: In the circuit diagram shown below, ammeter A₁ reads 10 amperes.

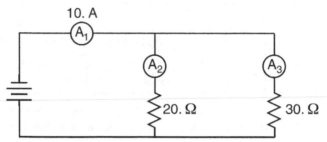

What is the reading of ammeter A₂?
(1) 6 A
(2) 10 A
(3) 20 A
(4) 4 A

9.40 A: (1) 6 A

	V	I	R	P
R₁	120V	6A	20Ω	720W
R₂	120V	4A	30Ω	480W
Total	120V	10A	12Ω	1200W

You can find more practice problems on the APlusPhysics website at: http://www.aplusphysics.com/regents.

Chapter 10: Magnetism

"Magnetism, as you recall from physics class, is a powerful force that causes certain items to be attracted to refrigerators."

— Dave Barry

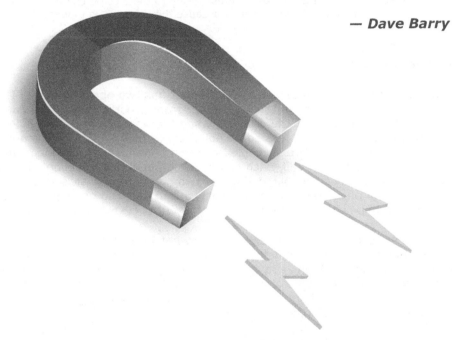

Objectives

1. Explain that magnetism is caused by moving electrical charges.
2. Describe the magnetic poles and interactions between magnets.
3. Draw magnetic field lines.
4. Describe the factors affecting an induced potential difference due to magnetic field lines interacting with moving charges.

Magnetism is closely related to electricity. In essence, **magnetism** is a force caused by moving charges. In the case of permanent magnets, the moving charges are the orbits of electrons spinning around nuclei. In very basic terms, strong permanent magnets have many atoms with electrons spinning in the same direction. Non-magnets have more random arrangements of electron spin around the nucleus. For electromagnets, the current itself provides the moving charges. In all cases, magnetic fields can be used to describe the forces due to magnets.

10.01 Q: Which type of field is present near a moving electric charge?

    (1) an electric field, only

    (2) a magnetic field, only

    (3) both an electric field and a magnetic field

    (4) neither an electric field nor a magnetic field

10.01 A: (3) An electric field is present due to the electric charge, and a magnetic field is present because the charge is in motion.

Magnetic Fields

Magnets are polarized, meaning every magnet has two opposite ends. The end of a magnet that points toward the geographic north pole of the Earth is called the north pole of the magnet, while the opposite end, for obvious reasons, is called the magnet's south pole. Every magnet has both a north and a south pole. There are no single isolated magnetic poles, or monopoles. If you split a magnet in half, each half of the original magnet then exhibits both a north and a south pole, giving you two magnets. Physicists continue to search both physically and theoretically, but to date, no one has ever observed a north pole without a south pole, or a south pole without a north pole.

We used electric field lines to help visualize what would happen to a positive charge placed in an electric field. In order to visualize a magnetic field, we can draw magnetic field lines (also known as magnetic flux lines) which show the direction the north pole of a magnet would tend to point if placed in the field. Magnetic field lines are drawn as closed loops, starting from the north pole of a magnet and continuing to the south pole of a magnet. Inside the magnet itself, the field lines run from the south pole to the north pole. The magnetic field is strongest in areas of greatest density of magnetic field lines, or areas of the greatest magnetic flux density.

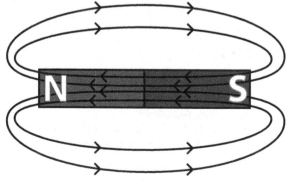

Much like electrical charges, like poles exert a repelling force on each other, while opposite poles exert an attractive force on each other. Materials can be classified as magnets, magnet attractables (materials which aren't magnets themselves but can be attracted by magnets), and non-attractables.

10.02 Q: The diagram below shows the lines of magnetic force between two north magnetic poles. At which point is the magnetic field strength greatest?

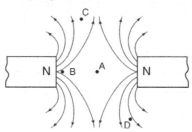

10.02 A: (B) has the greatest magnetic field strength because it is located at the highest density of magnetic field lines.

10.03 Q: The diagram below represents a 0.5-kilogram bar magnet and a 0.7-kilogram bar magnet with a distance of 0.2 meter between their centers.

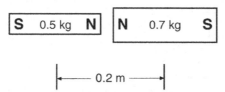

Which statement best describes the forces between the bar magnets?

(1) Gravitational force and magnetic force are both repulsive.

(2) Gravitational force is repulsive and magnetic force is attractive.

(3) Gravitational force is attractive and magnetic force is repulsive.

(4) Gravitational force and magnetic force are both attractive.

10.03 A: (3) Gravity always attracts and the north poles repel each other.

10.04 Q: A student is given two pieces of iron and told to determine if one or both of the pieces are magnets. First, the student touches an end of one piece to one end of the other. The two pieces of iron attract. Next, the student reverses one of the pieces and again touches the ends together. The two pieces attract again. What does the student definitely know about the initial magnetic properties of the two pieces of iron?

10.04 A: At least one of the pieces of iron is a magnet, but we cannot state with certainty that both are magnets.

10.05 Q: Draw a minimum of four field lines to show the magnitude and direction of the magnetic field in the region surrounding a bar magnet.

10.05 A:

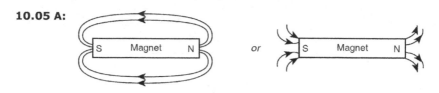

10.06 Q: When two ring magnets are placed on a pencil, magnet A remains suspended above magnet B, as shown below.

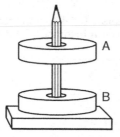

Which statement describes the gravitational force and the magnetic force acting on magnet A due to magnet B?

(1) The gravitational force is attractive and the magnetic force is repulsive.

(2) The gravitational force is repulsive and the magnetic force is attractive.

(3) Both the gravitational force and the magnetic force are attractive.

(4) Both the gravitational force and the magnetic force are repulsive.

10.06 A: (1) Gravity can only attract, and because magnet A is suspended above magnet B, the magnetic force must be repulsive.

The Compass

Because the Earth exerts a force on magnets (which, when used to tell direction, we call a compass), we can conclude that the Earth is a giant magnet. If the north pole of a magnet is attracted to the geographic north pole of the

Earth, and opposite poles attract, then it stands to reason that the geographic north pole of the Earth is actually a magnetic south pole! Compasses always line up with the net magnetic field.

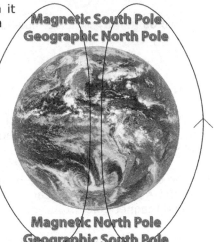

In truth, the magnetic north and south pole of the Earth are constantly moving. The current rate of change of the magnetic north pole is thought to be more than 20 kilometers per year, and it is believed that the magnetic north pole has shifted more than 1000 kilometers since it was first reached by explorer Sir John Ross in 1831!

10.07 Q: The diagram below represents the magnetic field near point P.

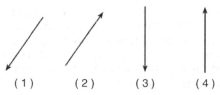

If a compass is placed at point P in the same plane as the magnetic field, which arrow represents the direction the north end of the compass needle will point?

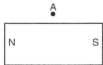

10.07 A: (2) Compass needles line up with the magnetic field.

10.08 Q: The diagram below shows a bar magnet.

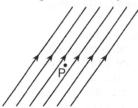

Which way will the needle of a compass placed at A point?

(1) up   (3) right

(2) down   (4) left

10.08 A: (3) since a compass lines up with the magnetic field.

Electromagnetism

In 1820, Danish physicist Hans Christian Oersted found that a current running through a wire created a magnetic field, kicking off the modern study of electromagnetism.

Moving electric charges create magnetic fields. You can test this by placing a compass near a current-carrying wire. The compass will line up with the induced magnetic field.

To determine the direction of the electrically-induced magnetic field, use the "right-hand-rule" by pointing your right-hand thumb in the direction of positive current flow. The curve of your fingers then shows the direction of the magnetic field around the wire.

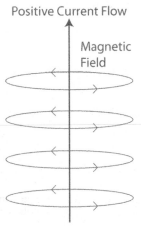

Not only do moving charges create magnetic fields, but relative motion between a conductor and a magnetic field can produce a potential difference in the conductor. The conductor must cut across the magnetic field lines to produce a potential difference, and larger potential differences are created when the conductor cuts across stronger magnetic fields, or moves more quickly through the magnetic field.

This phenomenon is what allows you to create usable, controllable electrical energy. Kinetic energy in the form of wind, water, steam, etc. is used to spin a coil of wire through a magnetic field, inducing a potential difference, which is transferred by the electric company to end users. This basic energy transformation is the underlying principle behind hydroelectric, nuclear, fossil fuel, and wind-powered electrical generators!

10.09 Q: The diagram below shows a wire moving to the right at speed v through a uniform magnetic field that is directed into the page. As the speed of the wire is increased, the induced potential difference will

Wire

X | X   X   X

v →

X | X   X   X

X | X   X   X

Magnetic field directed into page

(1) decrease

(2) increase

(3) remain the same

10.09 A: (2) the induced potential difference will increase as the speed of the wire is increased.

10.10 Q: The diagram below represents a wire conductor, RS, positioned perpendicular to a uniform magnetic field directed into the page.

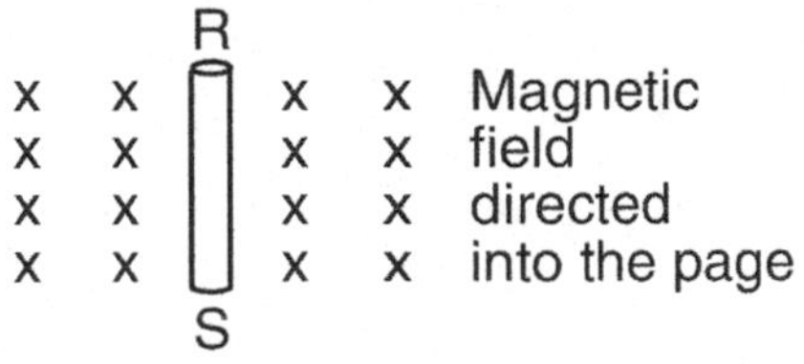

R

x x ⌐ x x Magnetic
x x | x x field
x x | x x directed
x x ⌐ x x into the page

S

Describe the direction in which the wire could be moved to produce the maximum potential difference across its ends, R and S.

10.10 A: The wire could be moved to produce the maximum potential difference across its ends, R and S, by moving it horizontally (right to left or left to right).

You can find more practice problems on the APlusPhysics website at: http://www.aplusphysics.com/regents.

Chapter 11: Waves

"It would be possible to describe everything scientifically, but it would make no sense; it would be without meaning, as if you described a Beethoven symphony as a variation of wave pressure."

— Albert Einstein

Objectives

1. Define a pulse.
2. Describe the behavior of a pulse at a boundary.
3. Understand how the principle of superposition is applied when two pulses meet.
4. Define three terms to describe periodic waves: speed, wavelength, and frequency.
5. Explain the characteristics of transverse and longitudinal waves.
6. Describe the formation of standing waves.
7. Apply the principle of superposition to the phenomenon of interference.
8. Understand how resonance occurs.
9. Understand the nature of sound waves.
10. Apply the Doppler effect qualitatively to problems involving moving sources or moving observers.
11. Explain the law of reflection.
12. Understand and apply Snell's law.
13. Calculate the index of refraction in a medium.
14. Relate the diffraction of light to its wave characteristics.
15. Describe Young's double-slit experiment.
16. Recognize characteristics of EM waves and determine the type of EM wave based on its characteristics.

Waves transfer energy through matter or space. We find waves everywhere: sound waves, light waves, microwaves, radio waves, water waves, earthquake waves, slinky waves, x-rays, and on and on. Developing an understanding of waves will allow us to understand how energy is transferred in the universe, and will eventually lead to a better understanding of matter and energy itself!

Wave Characteristics

A **pulse** is a single disturbance which carries energy through a medium or through space. Imagine you and your friend holding opposite ends of a slinky. If you quickly move your arm up and down, a single pulse will travel down the slinky toward your friend.

If, instead, you generate several pulses at regular time intervals, you now have a wave carrying energy down the slinky. A **wave**, therefore is a repeated disturbance which carries energy. The mass of the slinky doesn't move from one end of the slinky to the other, but the energy it carries does.

When a pulse or wave reaches a hard boundary, it reflects off the boundary, and is inverted. If a pulse or wave reaches a soft, or flexible, boundary, it still reflects off the boundary, but does not invert.

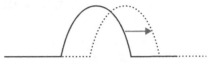

Waves can be classified in several different ways. One type of wave, known as a **mechanical wave**, requires a medium, or material, through which to travel. Examples of mechanical waves include water waves, sound waves, slinky waves, and even seismic waves. **Electromagnetic waves**, on the other hand, do not require a medium in order to travel. Electromagnetic waves (or EM waves) are considered part of the Electromagnetic Spectrum. Examples of EM waves include light, radio waves, microwaves, and even X-rays.

Further, waves can be classified based upon their direction of vibration. Waves in which the "particles" of the wave vibrate in the same direction as the wave velocity are known as **longitudinal**, or compressional, waves. Examples of longitudinal waves include sound waves and seismic P waves. Waves in which the particles of the wave vibrate perpendicular to the wave's direction of motion are known as **transverse** waves. Examples of transverse waves include seismic S waves, electromagnetic waves, and even stadium waves (the "human" waves you see at baseball and football games!).

Video animations of waves reflecting off boundaries as well as longitudinal and transverse waves can be viewed at http://bit.ly/gC1TMU.

Waves have a number of characteristics which define their behavior. Looking at a transverse wave, we can identify specific locations on the wave. The highest points on the wave are known as **crests**. The lowest points on the wave are known as **troughs**. The **amplitude** of the wave, corresponding to the energy of the wave, is the distance from the baseline to a crest or the baseline to a trough.

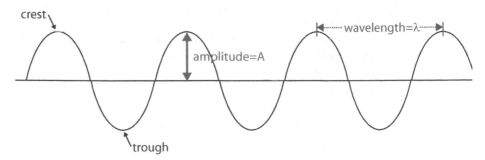

The length of the wave, or **wavelength**, represented by the Greek letter lambda (λ), is the distance between corresponding points on consecutive waves (i.e. crest to crest or trough to trough). Points on the same wave with the same displacement from equilibrium moving in the same direction (such as a crest to a crest or a trough to a trough) are said to be in phase (phase difference is 0° or 360°). Points with opposite displacements from equilibrium (such as a crest to a trough) are said to be 180° out of phase.

11.01 Q: Which type of wave requires a material medium through which to travel?

(1) sound

(2) television

(3) radio

(4) x ray

11.01 A: (1) sound is a mechanical wave and therefore requires a medium.

11.02 Q: The diagram below represents a transverse wave traveling to the right through a medium. Point A represents a particle of the medium.

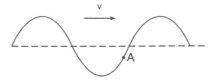

In which direction will particle A move in the next instant of time?

(1) up

(2) down

(3) left

(4) right

11.02 A: (2) particle A will move down as the wave passes.

11.03 Q: As a transverse wave travels through a medium, the individual particles of the medium move
(1) perpendicular to the direction of wave travel
(2) parallel to the direction of wave travel
(3) in circles
(4) in ellipses

11.03 A: (1) perpendicular to the direction of wave travel.

11.04 Q: A ringing bell is located in a chamber. When the air is removed from the chamber, why can the bell be seen vibrating but not be heard?
(1) Light waves can travel through a vacuum, but sound waves cannot.
(2) Sound waves have greater amplitude than light waves.
(3) Light waves travel slower than sound waves.
(4) Sound waves have higher frequency than light waves.

11.04 A: (1) Light is an EM wave, while sound is a mechanical wave.

11.05 Q: A single vibratory disturbance moving through a medium is called
(1) a node
(2) an antinode
(3) a standing wave
(4) a pulse

11.05 A: (4) a pulse.

11.06 Q: A periodic wave transfers
(1) energy, only
(2) mass, only
(3) both energy and mass
(4) neither energy nor mass

11.06 A: (1) energy, only.

11.07 Q: The diagram below represents a transverse wave.

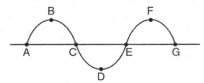

The wavelength of the wave is equal to the distance between points

(1) A and G

(2) B and F

(3) C and E

(4) D and F

11.07 A: (2) B and F is the wavelength as measured from crest to crest.

11.08 Q: The diagram below represents a periodic wave.

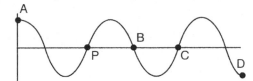

Which point on the wave is in phase with point P?

(1) A

(2) B

(3) C

(4) D

11.08 A: (3) Point C is the same point as point P but on a consecutive wave.

11.09 Q: The diagram below represents a transverse wave moving on a uniform rope with point A labeled. On the diagram, mark an X at the point on the wave that is 180° out of phase with point A.

11.09 A:

11.10 Q: The diagram below represents a transverse wave.

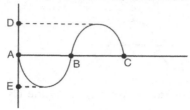

The distance between which two points identifies the amplitude of the wave?

(1) A and B

(2) A and C

(3) A and E

(4) D and E

11.10 A: (3) Amplitude is measured from the baseline to a crest or a trough, therefore amplitude is the distance between A and E.

11.11 Q: The diagram below represents a transverse wave traveling in a string.

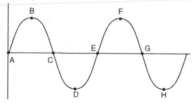

Which two labeled points are 180° out of phase?

(1) A and D

(2) B and F

(3) D and F

(4) D and H

11.11 A: (3) D and F.

In similar fashion, longitudinal waves also have amplitude and wavelength. In the case of longitudinal waves, however, instead of crests and troughs, the longitudinal waves have areas of high density (**compressions**) and areas of low density (**rarefactions**), as shown in the representation of the particles of a sound wave. The wavelength, then, of a compressional wave is the distance between compressions, or the distance between rarefactions. Once again, the amplitude corresponds to the energy of the wave.

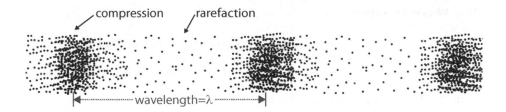

compression   rarefaction

|◄------------------- wavelength=λ -------------------►|

11.12 Q: A periodic wave is produced by a vibrating tuning fork. The amplitude of the wave would be greater if the tuning fork were

(1) struck more softly

(2) struck harder

(3) replaced by a lower frequency tuning fork

(4) replaced by a higher frequency tuning fork

11.12 A: (2) Striking the tuning fork harder gives the tuning fork more energy, increasing the sound wave's amplitude.

11.13 Q: Increasing the amplitude of a sound wave produces a sound with

(1) lower speed

(2) higher pitch

(3) shorter wavelength

(4) greater loudness

11.13 A: (4) greater loudness due to the greater energy / amplitude of the wave.

11.14 Q: A longitudinal wave moves to the right through a uniform medium, as shown below.

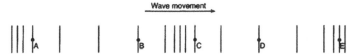

(A) Points A, B, C, D, and E represent the positions of particles of the medium. What is the direction of the motion of the particles at position C as the wave moves to the right?

(B) Between which two points on the wave could you measure a complete wavelength?

11.14 A: (A) The particles move to the left and right at position C, as the particles in a longitudinal wave vibrate parallel to the wave velocity.

(B) You could measure a complete wavelength between points A and C, since A and C represent the same point on successive waves.

The Wave Equation

The frequency (f) of a wave describes the number of waves that pass a given point in a time period of one second. The higher the frequency, the more waves that pass. Frequency is measured in number of waves per second (1/s), also known as a Hertz (Hz). If 60 waves pass a given point in a second, the frequency of the wave would be 60 Hz.

Closely related to frequency, the period (T) of a wave describes how long it takes for a single wave to pass a given point and can be found as the reciprocal of the frequency. Period is a measurement of time, and therefore is measured in seconds. Both frequency and period were introduced earlier in our discussion of circular motion.

11.15 Q: What is the period of a 60-hertz electromagnetic wave traveling at 3.0×10^8 meters per second?

11.15 A: $T = \dfrac{1}{f} = \dfrac{1}{60 Hz} = 0.0167s$

11.16 Q: Which unit is equivalent to meters per second?
(1) Hz•s
(2) Hz•m
(3) s/Hz
(4) m/Hz

11.16 A: (2) $\dfrac{m}{s} = Hz \bullet m$

11.17 Q: The product of a wave's frequency and its period is
(1) one
(2) its velocity
(3) its wavelength
(4) Planck's constant

11.17 A: (1) $f \bullet T = f \bullet \dfrac{1}{f} = 1$

Because waves move through space, they must have a velocity. The velocity of a wave is a function of the type of wave, and the medium it travels through. Electromagnetic waves moving through a vacuum, for instance, travel at roughly $3*10^8$ m/s. This value is so famous and important in physics it is given its own symbol, c. When an electromagnetic wave enters a different medium, such as glass, it slows down. If the same wave were to then re-emerge from glass back into a vacuum, it would again travel at c, or $3*10^8$ m/s.

The speed of a wave can be easily related to its frequency and wavelength. Speed of a wave is determined by the wave type and the medium it is traveling through. For a given wave speed, as frequency increases, wavelength must decrease, and vice versa. This can be shown mathematically using the wave equation.

$$v = f\lambda$$

11.18 Q: A periodic wave having a frequency of 5 hertz and a speed of 10 meters per second has a wavelength of

(1) 0.50 m

(2) 2.0 m

(3) 5.0 m

(4) 50. m

11.18 A: (2) $v = f\lambda$

$$\lambda = \frac{v}{f} = \frac{10\,^{m}/_{s}}{5\,Hz} = 2m$$

11.19 Q: If the amplitude of a wave is increased, the frequency of the wave will

(1) decrease

(2) increase

(3) remain the same

11.19 A: (3) remain the same.

11.20 Q: An electromagnetic wave traveling through a vacuum has a wavelength of 1.5×10^{-1} meters. What is the period of this electromagnetic wave?

(1) 5.0×10^{-10} s

(2) 1.5×10^{-1} s

(3) 4.5×10^{7} s

(4) 2.0×10^{9} s

11.20 A: (1) $v = f\lambda = \dfrac{\lambda}{T}$

$$T = \frac{\lambda}{v} = \frac{1.5 \times 10^{-1}\,m}{3 \times 10^{8}\,\frac{m}{s}} = 5 \times 10^{-10}\,s$$

11.21 Q: A surfacing blue whale produces water wave crests having an amplitude of 1.2 meters every 0.40 seconds. If the water wave travels at 4.5 meters per second, the wavelength of the wave is

(1) 1.8 m

(2) 2.4 m

(3) 3.0 m

(4) 11 m

11.21 A: (1) $v = f\lambda$

$$\lambda = \frac{v}{f} = vT = (4.5\,\tfrac{m}{s})(0.4s) = 1.8m$$

Sound Waves

Sound is a mechanical wave which we observe by detecting vibrations in the inner ear. Typically, we think of sound as traveling through air, therefore the particles vibrating are air molecules. Sound can travel through other media as well, including water, wood, and even steel.

The particles of a sound wave vibrate in a direction parallel with the direction of the sound wave's velocity, therefore sound is a longitudinal wave. The speed of sound in air at standard temperature and pressure (STP) is 331 m/s, a value which is supplied on the front of the Regents Physics Reference Table.

11.22 Q: At an outdoor physics demonstration, a delay of 0.50 seconds was observed between the time sound waves left a loudspeaker and the time these sound waves reached a student through the air. If the air is at STP, how far was the student from the speaker?

11.22 A: $\bar{v} = \dfrac{d}{t} \Rightarrow d = \bar{v}t = (331\tfrac{m}{s})(0.50s) = 166m$

11.23 Q: The sound wave produced by a trumpet has a frequency of 440 hertz. What is the distance between successive compressions in this sound wave as it travels through air at STP?

(1) 1.5×10^{-6} m

(2) 0.75 m

(3) 1.3 m

(4) 6.8×10^{5} m

11.23 A: (2) $v = f\lambda$

$$\lambda = \frac{v}{f} = \frac{331\,m/_s}{440\,Hz} = 0.75m$$

11.24 Q: A stationary research ship uses sonar to send a 1.18×10^{3}-hertz sound wave down through the ocean water. The reflected sound wave from the flat ocean bottom 324 meters below the ship is detected 0.425s second after it was sent from the ship.

(A) Calculate the speed of the sound wave in the ocean water.

(B) Calculate the wavelength of the sound wave in the water.

(C) Determine the period of the sound wave in the water.

11.24 A: (A) $\overline{v} = \frac{d}{t} = \frac{648m}{0.425s} = 1520\,m/_s$

(B) $v = f\lambda$

$$\lambda = \frac{v}{f} = \frac{1525\,m/_s}{1180\,Hz} = 1.29m$$

(C) $T = \frac{1}{f} = \frac{1}{1180\,Hz} = 8.47 \times 10^{-4}s$

When we observe sound waves through hearing, we pick up the amplitude, or energy, of the waves as loudness. The frequency of the wave is perceived as pitch, with higher frequencies observed as a higher pitch. Typically, humans can hear a frequency range of 20Hz to 20,000 Hz, although young observers can often detect frequencies above 20,000 Hz, an ability which declines with age.

Certain devices create strong sound waves at a single specific frequency. If another object, having the same "**natural frequency**," is impacted by these sound waves, it may begin to vibrate at this frequency, producing more sound waves. The phenomenon where one object emitting a sound wave with a specific frequency causes another object with the same natural frequency to vibrate is known as **resonance**. A dramatic demonstration of resonance involves an opera singer breaking a glass by singing a high pitch note. The singer creates a sound wave with a frequency equal to the natural frequency of the glass, causing the glass to vibrate at its natural, or resonant, frequency so energetically that it shatters.

11.25 Q: Sound waves strike a glass and cause it to shatter. This phenomenon illustrates

(1) resonance

(2) refraction

(3) reflection

(4) diffraction

11.25 A: (1) resonance

11.26 Q: A dampened fingertip rubbed around the rim of a crystal glass causes the glass to vibrate and produce a musical note. This effect is due to

(1) resonance

(2) refraction

(3) reflection

(4) rarefaction

11.26 A: (1) resonance

11.27 Q: Resonance occurs when one vibrating object transfers energy to a second object causing it to vibrate. The energy transfer is most efficient when, compared to the first object, the second object has the same natural

(1) frequency

(2) loudness

(3) amplitude

(4) speed

11.27 A: (1) frequency.

11.28 Q: A car traveling at 70 kilometers per hour accelerates to pass another car. When the car reaches a speed of 90 kilometers per hour the driver hears the glove compartment door start to vibrate. By the time the speed of the car is 100 kilometers per hour, the glove compartment door has stopped vibrating. This vibrating phenomenon is an example of

(1) destructive interference

(2) the Doppler effect

(3) diffraction

(4) resonance

11.28 A: (4) resonance.

11.29 Q: Which wave phenomenon occurs when vibrations in one object cause vibrations in a second object?

(1) reflection

(2) resonance

(3) intensity

(4) tuning

11.29 A: (2) resonance.

Interference

When more than one wave travels through the same location in the same medium at the same time, the total displacement of the medium is governed by the principle of **superposition**. The principle of superposition simply states that the total displacement is the sum of all the individual displacements of the waves. The combined effect of the interaction of the multiple waves is known as **wave interference**.

11.30 Q: The diagram below shows two pulses approaching each other in a uniform medium. Diagram the superposition of the two pulses.

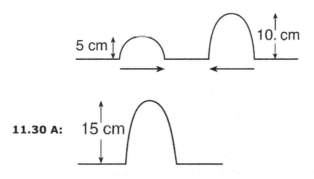

When two or more pulses with displacements in the same direction interact, the effect is known as **constructive interference**. The resulting displacement is greater than the original individual pulses. Once the pulses have passed by each other, they continue along their original path in their original shape, as if they had never met.

When two or more pulses with displacements in opposite directions interact, the effect is known as **destructive interference**. The resulting displacements negate each other. Once the pulses have passed by each other, they continue along their original path in their original shape, as if they had never met. An animation of two pulses interfering constructively and destructively is available at http://bit.ly/hyJ3lZ.

11.31 Q: The diagram below represents two pulses approaching each other from opposite directions in the same medium.

Which diagram best represents the medium after the pulses have passed through each other?

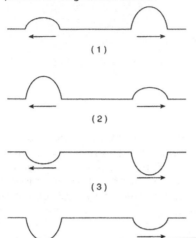

11.31 A: (2) the pulses continue as if they had never met.

11.32 Q: The diagram below represents shallow water waves of constant wavelength passing through two small openings, A and B, in a barrier.

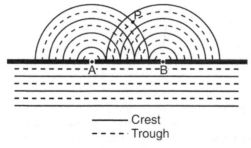

——— Crest
- - - - Trough

Which statement best describes the interference at point P?

(1) It is constructive, and causes a longer wavelength.

(2) It is constructive, and causes an increase in amplitude.

(3) It is destructive, and causes a shorter wavelength.

(4) It is destructive, and causes a decrease in amplitude.

11.32 A: (4) when a crest and a trough meet, destructive interference causes a decrease in amplitude.

11.33 Q: The diagram below shows two pulses of equal amplitude, A, approaching point P along a uniform string.

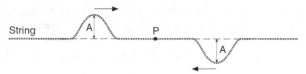

When the two pulses meet at P, the vertical displacement of the string at P will be

(1) A

(2) 2A

(3) 0

(4) A/2

11.33 A: (3) the pulses will experience destructive interference.

11.34 Q: The diagram below represents two pulses approaching each other.

Which diagram best represents the resultant pulse at the instant the pulses are passing through each other?

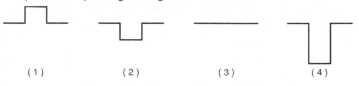

(1)   (2)   (3)   (4)

11.34 A: (2) shows the superposition (addition) of the two pulses.

11.35 Q: Two waves having the same amplitude and frequency are traveling in the same medium. Maximum destructive interference will occur when the phase difference between the waves is

(1) 0°

(2) 90°

(3) 180°

(4) 270°

11.35 A: (3) Maximum destructive interference occurs at a phase difference of 180°.

Standing Waves

When waves of the same frequency and amplitude traveling in opposite directions meet, a standing wave is produced. A **standing wave** is a wave in which certain points (**nodes**) appear to be standing still and other points (**antinodes**) vibrate with maximum amplitude above and below the axis.

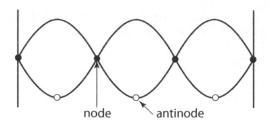

node   antinode

Looking at the standing wave produced above, we can see a total of four nodes in the wave, and three antinodes. For any standing wave pattern, you will always have one more node than antinode.

Standing waves can be observed in a variety of patterns and configurations, and are responsible for the functioning of most musical instruments. Guitar strings, for example, demonstrate a standing wave pattern. By fretting the strings, you adjust the wavelength of the string, and therefore the frequency of the standing wave pattern, creating a different pitch. Similar functionality is seen in instruments ranging from pianos and drums to flutes, harps, trombones, xylophones, and even pipe organs!

11.36 Q: While playing, two children create a standing wave in a rope, as shown in the diagram below.

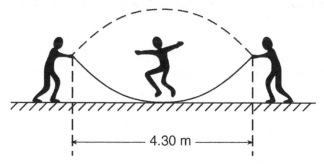

|← —————— 4.30 m —————— →|

A third child participates by jumping the rope. What is the wavelength of this standing wave?

(1) 2.15 m

(2) 4.30 m

(3) 6.45 m

(4) 8.60 m

11.36 A: (4) the standing wave shown is half a wavelength, therefore the total wavelength must be 8.6m.

11.37 Q: Wave X travels eastward with frequency f and amplitude A. Wave Y, traveling in the same medium, interacts with wave X and produces a standing wave. Which statement about wave Y is correct?

(1) Wave Y must have a frequency of f, an amplitude of A, and be traveling eastward.

(2) Wave Y must have a frequency of 2f, an amplitude of 3A, and be traveling eastward.

(3) Wave Y must have a frequency of 3f, an amplitude of 2A, and be traveling westward.

(4) Wave Y must have a frequency of f, an amplitude of A, and be traveling westward.

11.37 A: (4) Standing waves are created when waves with the same frequency and amplitude traveling in opposite directions meet.

11.38 Q: The diagram below represents a wave moving toward the right side of this page.

Which wave shown below could produce a standing wave with the original wave?

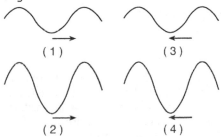

11.38 A: (3) must have same frequency, amplitude, and be traveling in the opposite direction in the same medium.

11.39 Q: The diagram below shows a standing wave.

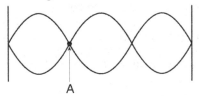

A

Point A on the standing wave is

(1) a node resulting from constructive interference
(2) a node resulting from destructive interference
(3) an antinode resulting from constructive interference
(4) an antinode resulting from destructive interference

11.39 A: (2) a node resulting from destructive interference.

11.40 Q: One end of a rope is attached to a variable speed drill and the other end is attached to a 5-kilogram mass. The rope is draped over a hook on a wall opposite the drill. When the drill rotates at a frequency of 20 Hz, standing waves of the same frequency are set up in the rope. The diagram below shows such a wave pattern.

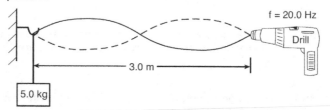

(A) Determine the wavelength of the waves producing the standing wave pattern.

(B) Calculate the speed of the wave in the rope.

11.40 A: (A) Wavelength is 3.0 meters from diagram.

(B) $v = f\lambda = (20\,Hz)(3m) = 60\,{}^m\!/_s$

Due to their very nature, waves exhibit a number of behaviors that may not be obvious upon first inspection, including the Doppler Effect, reflection, refraction, and diffraction. Understanding these behaviors brings us closer to understanding the universe, while also providing a number of useful applications including, but not limited to, radar, sonography, digital televisions, mirrors, telescopes, glasses, contact lenses, atomic research, and even holography!

Doppler Effect

The shift in a wave's observed frequency due to relative motion between the source of the wave and the observer is known as the **Doppler Effect**. In essence, when the source and/or observer are moving toward each other, the observer perceives a shift to a higher frequency, and when the source and/or observer are moving away from each other, the observer perceives a lower frequency.

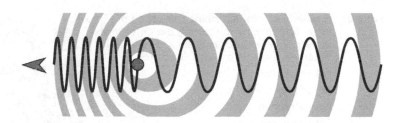

This can be observed when a vehicle travels past you. As you hear the vehicle approach, you can observe a higher frequency noise, and as the vehicle passes by you and then moves away, you observe a lower frequency noise. This effect is the principle behind radar guns to measure an object's speed as well as meteorology radar which provides data on wind speeds.

The Doppler Effect results from waves having a fixed speed in a given medium. As waves are emitted, a moving source or observer encounters the wave fronts at a different frequency than the waves are emitted, resulting in a perceived shift in frequency. The video and animation at http://bit.ly/epLkPj may help you visualize this effect.

11.41 Q: A car's horn produces a sound wave of constant frequency. As the car speeds up going away from a stationary spectator, the sound wave detected by the spectator

(1) decreases in amplitude and decreases in frequency

(2) decreases in amplitude and increases in frequency

(3) increases in amplitude and decreases in frequency

(4) increases in amplitude and increases in frequency

11.41 A: (1) decreases in amplitude because the distance between source and observe is increasing, and decreases in frequency because the source is moving away from the observer.

11.42 Q: A car's horn is producing a sound wave having a constant frequency of 350 hertz. If the car moves toward a stationary observer at constant speed, the frequency of the car's horn detected by this observer may be

(1) 320 Hz

(2) 330 Hz

(3) 350 Hz

(4) 380 Hz

11.42 A: (4) If source is moving toward the stationary observer, the observed frequency must be higher than source frequency.

11.43 Q: A radar gun can determine the speed of a moving automobile by measuring the difference in frequency between emitted and reflected radar waves. This process illustrates

(1) resonance

(2) the Doppler effect

(3) diffraction

(4) refraction

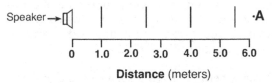

11.43 A: (2) the Doppler effect.

11.44 Q: The vertical lines in the diagram represent compressions in a sound wave of constant frequency propagating to the right from a speaker toward an observer at point A.

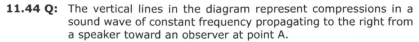

(A) Determine the wavelength of this sound wave.

(B) The speaker is then moved at constant speed toward the observer at A. Compare the wavelength of the sound wave received by the observer while the speaker is moving to the wavelength observed when the speaker was at rest.

11.44 A: (A) Wavelength is compression to compression, or 1.5m.

(B) Observed frequency is higher while speaker is moving toward the observer due to the Doppler Effect, so the observed wavelength must be shorter.

11.45 Q: A student sees a train that is moving away from her and sounding its whistle at a constant frequency. Compared to the sound produced by the whistle, the sound observed by the student is

(1) greater in amplitude

(2) a transverse wave rather than a longitudinal wave

(3) higher in pitch

(4) lower in pitch

11.45 A: (4) lower in pitch since the source is moving away from the observer.

An exciting application of the Doppler Effect involves the analysis of radiation from distant stars and galaxies in the universe. Based on the basic elements that compose stars, we know what frequencies of radiation to look for. However, when analyzing these objects, we observe frequencies shifted toward the red end of the electromagnetic spectrum (lower frequencies), known as the **Red Shift**. This indicates that these celestial objects must be moving away from us. The more distant the object, the greater the red shift. Putting this together, we can conclude that more distant celestial objects are moving away from us faster, and therefore, the universe as we know it must be expanding!

11.46 Q: When observed from Earth, the wavelengths of light emitted by a star are shifted toward the red end of the electromagnetic spectrum. This redshift occurs because the star is

(1) at rest relative to Earth

(2) moving away from Earth

(3) moving toward Earth at decreasing speed

(4) moving toward Earth at increasing speed

11.46 A: (2) moving away from Earth.

Reflection

When a wave hits a boundary, three different events can occur. The wave may be:

- Reflected - wave bounces off a boundary
- Transmitted - wave is transmitted into the new medium
- Absorbed - energy of the wave is transferred into the boundary medium

The **law of reflection** states that the angle at which a wave strikes a reflective medium (the **angle of incidence**, or θ_i) is equal to the angle at which a wave reflects off the medium (the **angle of reflection**, or θ_r). Put more simply, $\theta_i = \theta_r$. In all cases, the angle of incidence and the angle of reflection are measured from a line perpendicular, or normal, to the reflecting surface.

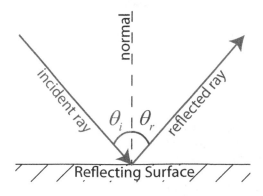

Although all waves can exhibit these behaviors, electromagnetic light waves are typically considered for demonstration purposes. When a wave bounces off a reflective surface, the nature of its reflection depends largely on the nature of the surface. Rough surfaces tend to reflect light in a variety of directions in a process known as diffuse reflection. **Diffuse reflection** is the type of reflection typically observed off of pieces of paper. Smooth surfaces tend to reflect light waves in a more regular fashion, such that the reflected rays maintain their parallelism. This process is known as **specular reflection**, and is commonly observed in mirrors.

11.47 Q: The diagram below represents a light ray striking the boundary between air and glass.

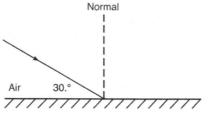

What would be the angle between this light ray and its reflected ray?

(1) 30°

(2) 60°

(3) 120°

(4) 150°

11.47 A: (3) recall that ray angles are always measured to the normal, so the angle between the two rays is 60°+60°=120°.

11.48 Q: A sonar wave is reflected from the ocean floor. For which angles of incidence do the wave's angle of reflection equal its angle of incidence?

(1) angles less than 45°, only

(2) an angle of 45°, only

(3) angles greater than 45°, only

(4) all angles of incidence

11.48 A: (4) the law of reflection applies to all types of waves reflecting off a surface.

Refraction

When a wave reaches the boundary between media, part of the wave is reflected and part of the wave enters the new medium. As the wave enters the new medium, the speed of the wave changes, and the frequency of a wave remains constant, therefore, consistent with the wave equation, $v=f\lambda$, the wavelength of the wave must change.

11.49 Q: When a wave enters a new material, what happens to its speed, frequency, and wavelength?

11.49 A: Speed changes, frequency remains constant, and wavelength changes.

The front of a wave has some actual width, and if the wave does not impinge upon the boundary between media at a right angle, not all of the wave enters the new medium and changes speed at the same time. This causes the wave to bend as it enters a new medium in a process known as **refraction**.

To better illustrate this, imagine you're in a line in a marching band, connected with your bandmates as you march at a constant speed down the field in imitation of a wave front. As your wavefront reaches a new medium that slows you down, such as a mud pit, the band members reaching the mud pit slow down before those who reach the pit later. Since you are all connected in a wave front, the entire wave shifts directions (refracts) as it passes through the boundary between field and mud!

The **index of refraction** (n) is a measure of how much light slows down in a material. In a vacuum and in air, all electromagnetic waves have a speed of $c=3\times10^8$ m/s. In other materials, light slows down. The ratio of the speed of light in a vacuum to the speed of light in the new material is known as the index of refraction (n). The slower the wave moves in the material, the larger the index of refraction:

$$n = \frac{c}{v}$$

Not only does index of refraction depend upon the medium the light wave is traveling through, it also varies with frequency. Thankfully, its variation is typically fairly small, and the Regents Physics Reference Table even provides you a table of indices of refraction for common materials at a set frequency.

Absolute Indices of Refraction	
$(f = 5.09 \times 10^{14}$ Hz)	
Air	1.00
Corn oil	1.47
Diamond	2.42
Ethyl alcohol	1.36
Glass, crown	1.52
Glass, flint	1.66
Glycerol	1.47
Lucite	1.50
Quartz, fused	1.46
Sodium chloride	1.54
Water	1.33
Zircon	1.92

11.50 Q: A light ray traveling in air enters a second medium and its speed slows to 1.71×10^8 m/s. What is the absolute index of refraction of the second medium?

11.50 A: $n = \dfrac{c}{v} = \dfrac{3 \times 10^8 \, m/s}{1.71 \times 10^8 \, m/s} = 1.75$

11.51 Q: In which way does blue light change as it travels from diamond into crown glass?

(1) Its frequency decreases.

(2) Its frequency increases.

(3) Its speed decreases.

(4) Its speed increases.

11.51 A: (4) Its speed increases because it crosses from a higher index of refraction material to a lower index of refraction material.

11.52 Q: Which characteristic is the same for every color of light in a vacuum?

(1) energy

(2) frequency

(3) speed

(4) period

11.52 A: (3) the speed of all EM waves in a vacuum is 3.0×10^8 m/s.

11.53 Q: A periodic wave travels at speed v through medium A. The wave passes with all its energy into medium B. The speed of the wave through medium B is v/2. Draw the wave as it travels through medium B.

11.53 A:

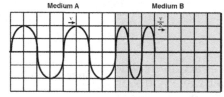

11.54 Q: A beam of monochromatic light has a wavelength of 5.89×10^{-7} meters in air. Calculate the wavelength of this light in diamond.

11.54 A:
$$\frac{n_2}{n_1} = \frac{\lambda_1}{\lambda_2}$$

$$\lambda_2 = \frac{n_1 \lambda_1}{n_2} = \frac{(1.00)(5.89 \times 10^{-7}\, m)}{2.42} = 2.43 \times 10^{-7}\, m$$

11.55 Q: The speed of light in a piece of plastic is 2.00×10^8 meters per second. What is the absolute index of refraction of this plastic?

(1) 1.00
(2) 0.670
(3) 1.33
(4) 1.50

11.55 A: (4) $n = \dfrac{c}{v} = \dfrac{3 \times 10^8 \; ^m\!/_s}{2 \times 10^8 \; ^m\!/_s} = 1.5$

The amount a light wave bends as it enters a new medium is given by the law of refraction, also known as **Snell's Law**, which states:

$$n_1 \sin \theta_1 = n_2 \sin \theta_2$$

In this formula, n_1 and n_2 are the indices of refraction of the two media, and θ_1 and θ_2 correspond to the angles of the incident and refracted rays, again measured from the normal. Light bends toward the normal as it enters a material with a higher index of refraction (slower material), and bends away from the normal as it enters a material with a lower index of refraction (faster material).

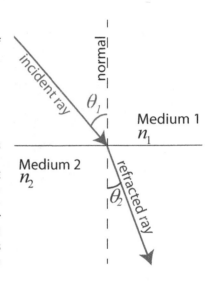

11.56 Q: A ray of light ($f=5.09 \times 10^{14}$ Hz) traveling in air is incident at an angle of 40° on an air-crown glass interface as shown below.

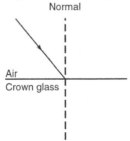

What is the angle of refraction for this light ray?

(1) 25°
(2) 37°
(3) 40°
(4) 78°

11.56 A: (1) $n_1 \sin\theta_1 = n_2 \sin\theta_2$

$$\theta_2 = \sin^{-1}\left(\frac{n_1 \sin\theta_1}{n_2}\right) = \sin^{-1}\left(\frac{1.00 \times \sin 40°}{1.52}\right) = 25°$$

11.57 Q: A ray of monochromatic light (f =5.09×10¹⁴ Hz) passes from air into Lucite at an angle of incidence of 30°.

(A) Calculate the angle of refraction in the Lucite.

(B) Using a protractor and straightedge, draw the refracted ray in the Lucite.

11.57 A: (A) $n_1 \sin\theta_1 = n_2 \sin\theta_2$

$$\theta_2 = \sin^{-1}\left(\frac{n_1 \sin\theta_1}{n_2}\right) = \sin^{-1}\left(\frac{1.00 \times \sin 30°}{1.50}\right) = 19°$$

(B)

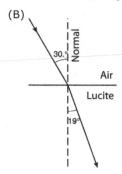

11.58 Q: When a light wave enters a new medium and is refracted, there must be a change in the light wave's

(1) color

(2) frequency

(3) period

(4) speed

11.58 A: (4) the change in a wave's speed causes its refraction.

11.59 Q: A ray of light (f=5.09×10¹⁴ Hz) traveling in air strikes a block of sodium chloride at an angle of incidence of 30°. What is the angle of refraction for the light ray in the sodium chloride?

(1) 19°

(2) 25°

(3) 40°

(4) 49°

11.59 A: (1) $n_1 \sin \theta_1 = n_2 \sin \theta_2$

$$\theta_2 = \sin^{-1}\left(\frac{n_1 \sin \theta_1}{n_2} \right) = \sin^{-1}\left(\frac{1.00 \times \sin 30°}{1.54} \right) = 19°$$

11.60 Q: Which diagram best represents the behavior of a ray of mono-chromatic light in air incident on a block of crown glass?

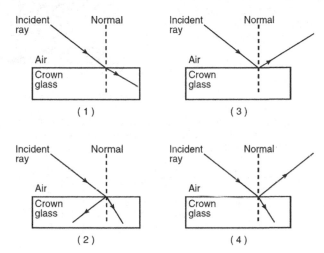

11.60 A: (4) shows both reflection and refraction of the incoming light ray.

Diffraction

Diffraction is the bending of waves around obstacles, or the spreading of waves as they pass through an opening, most apparent when looking at obstacles or wavelengths having a size of the same order of magnitude as the wavelength. Typically, the smaller the obstacle and longer the wavelength, the greater the diffraction. Taken to the extreme, when a wave is blocked by a small enough opening, the wave passing through the opening actually behaves like a point source for a new wave.

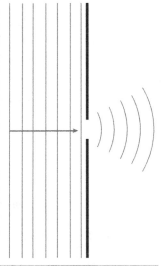

You can observe diffraction quite easily. I'm sure you've heard a noise from a room with an open door even when your ears aren't in a direct line from the sound source. This is a result of diffraction of the sound waves around the door opening (along with some reflection of sound as well).

Thomas Young's Double-Slit Experiment is a famous experiment which utilized diffraction to prove light has properties of waves. Young placed a single-wavelength light source behind a barrier with two narrow slits, allowing only a small portion of the light to pass through each slit. Because the two light waves travel different distances to the screen on which they are projected, you can see effects of both constructive and destructive interference, phenomena that occur only for waves!

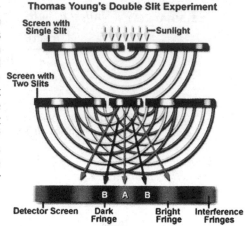

Thomas Young's Double Slit Experiment

Illustration Courtesy of Michael W. Davidson

11.61 Q: Which diagram best represents the shape and direction of a series of wave fronts after they have passed through a small opening in a barrier?

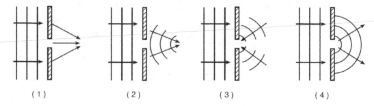

(1)   (2)   (3)   (4)

11.61 A: (4) the wave spreads out as it passes through a small opening.

11.62 Q: A beam of monochromatic light approaches a barrier having four openings, A, B, C, and D, of different sizes as shown below.

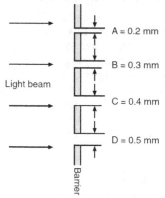

A = 0.2 mm

B = 0.3 mm

Light beam

C = 0.4 mm

D = 0.5 mm

Which opening will cause the greatest diffraction?

11.62 A: (A) has the smallest opening, so will create the most diffraction.

11.63 Q: Parallel wave fronts incident on an opening in a barrier are dif-
fracted. For which combination of wavelength and size of opening will diffraction effects be greatest?

(1) short wavelength and narrow opening

(2) short wavelength and wide opening

(3) long wavelength and narrow opening

(4) long wavelength and wide opening

11.63 A: (3) long wavelength and narrow opening produces the greatest
diffraction.

11.64 Q: A wave of constant wavelength diffracts as it passes through an
opening in a barrier. As the size of the opening is increased, the
diffraction effects

(1) decrease

(2) increase

(3) remain the same

11.64 A: (1) As the size of the opening increases, the amount of diffraction decreases.

11.65 Q: The diagram below shows a plane wave passing through a small
opening in a barrier.

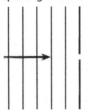

Sketch four wave fronts after they have passed through the barrier.

11.65 A:

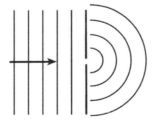

Electromagnetic Spectrum

Unlike mechanical waves, electromagnetic (EM) waves do not require a medium in which to travel. They consist of an electric field component and a magnetic field component oriented perpendicular to each other and to the wave velocity, and are caused by vibrating electrical charges. The orientation of the electric field and magnetic field components of an electromagnetic wave can be visualized below.

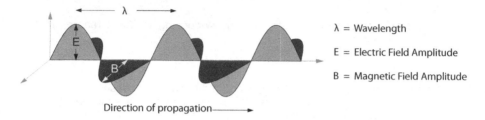

λ = Wavelength

E = Electric Field Amplitude

B = Magnetic Field Amplitude

Direction of propagation

The speed of all electromagnetic waves in a vacuum is approximately 3×10^8 m/s. This constant is so important in physics that it is represented by the letter c, and is, according to our current understanding of the universe, the fastest possible speed anything in the universe can travel.

Since c is a constant for all EM waves in a vacuum, the product of frequency and wavelength must be a constant. Therefore, at higher frequencies, EM waves have a shorter wavelength, and at lower frequencies, EM waves have a longer wavelength. If the EM wave travels into a new medium, its speed can decrease (see page 210), and because frequency remains constant, its wavelength would also decrease.

It is the frequency of an EM wave that determines its characteristics. The relationship between frequency and wavelength in a vacuum for various types of EM waves is depicted in the Electromagnetic Spectrum, provided in the Regents Physics Reference Table. This diagram can be useful for answering questions and solving problems involving electromagnetic waves.

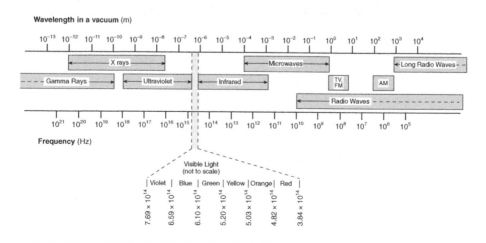

The electromagnetic spectrum describes the types of electromagnetic waves observed at the specified frequencies and wavelength. It is also important to note that the energy of an electromagnetic wave is directly related to its frequency, therefore higher frequency (shorter wavelength) EM waves have more energy than lower frequency (longer wavelength) EM waves.

An x-ray, therefore, has considerably more energy than an AM radio wave! Using the diagram, more energetic waves are shown on the left side of the EM Spectrum, and less energetic waves are shown to the right on the EM Spectrum. We'll explore the energy of EM radiation further in the Modern Physics chapter.

11.66 Q: Which color of light has a wavelength of 5.0×10^{-7} meters in air?

(1) blue

(2) green

(3) orange

(4) violet

11.66 A: (2) First find the frequency using v=fλ, then use the Electromagnetic Spectrum to determine the correct color based on the frequency.

$$v = f\lambda$$

$$f = \frac{v}{\lambda} = \frac{c}{\lambda} = \frac{3 \times 10^8 \, ^m\!/_s}{5.0 \times 10^{-7} \, m} = 6 \times 10^{14} \, Hz$$

11.67 Q: A television remote control is used to direct pulses of electromagnetic radiation to a receiver on a television. This communication from the remote control to the television illustrates that electromagnetic radiation

(1) is a longitudinal wave

(2) possesses energy inversely proportional to its frequency

(3) diffracts and accelerates in air

(4) transfers energy without transferring mass

11.67 A: (4) transfers energy without transferring mass.

11.68 Q: A microwave and an x ray are traveling in a vacuum. Compared to the wavelength and period of the microwave, the x ray has a wavelength that is

(1) longer and a period that is shorter

(2) longer and a period that is longer

(3) shorter and a period that is longer

(4) shorter and a period that is shorter

11.68 A: (4) shorter and a period that is shorter.

11.69 Q: A 1.50×10^{-6}-meter-long segment of an electromagnetic wave having a frequency of 6×10^{14} hertz is represented below.

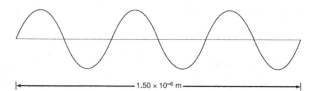

(A) Mark two points on the wave that are in phase with each other. Label each point with the letter P.

(B) According to the Reference Tables for Physical Setting/Physics, which type of electromagnetic wave does the segment in the diagram represent?

11.69 A: (A)

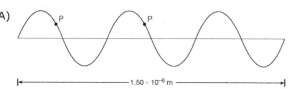

(B) green light (visible light).

11.70 Q: What is the period of a 60-hertz electromagnetic wave traveling at 3.0×10^8 meters per second?

(1) 1.7×10^{-2} s

(2) 2.0×10^{-7} s

(3) 6.0×10^1 s

(4) 5.0×10^6 s

11.70 A: (1) $T = \dfrac{1}{f} = \dfrac{1}{60Hz} = 0.017s$

You can find more practice problems on the APlusPhysics website at: http://www.aplusphysics.com/regents.

Chapter 12: Modern Physics

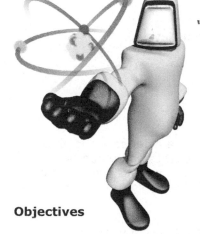

"God does not play dice with the cosmos."
— Albert Einstein

"Einstein, don't tell God what to do."
— Niels Bohr

Objectives

1. Explain the wave-particle duality of light.
2. Calculate the energy of a photon from its wave characteristics.
3. Calculate the energy of an absorbed or emitted photon from an energy level diagram.
4. Explain the quantum nature of atomic energy levels.
5. Explain the Rutherford and Bohr models of the atom.
6. Explain the universal conservation laws.
7. Recognize the fundamental source of all energy in the universe as the conversion of mass into energy.
8. Understand and use the mass-energy equivalence equation.
9. Understand that atomic particles are composed of subnuclear particles.
10. Explain how the nucleus is a conglomeration of quarks which combine to form protons and neutrons.
11. Understand that each elementary particle has a corresponding anti-particle.
12. Utilize Standard Model diagrams to solve basic particle physics problems.
13. Define the known fundamental forces in the universe and rank them in order of relative strength.

Modern Physics refers largely to advancements in physics from the 1900s to the present, extending our models of Newtonian (classical) mechanics and electricity and magnetism to the extremes of the very small, the very large, the very slow and the very fast. Modern Physics can encompass a tremendous variety of topics, which we will briefly explore in this book. Key topics for our exploration include:

- models of the atom
- sub-atomic structure
- universal conservation laws
- mass-energy equivalence
- fundamental forces in the universe
- the dual nature of electromagnetic radiation
- the quantum nature of atomic energy levels

Wave-Particle Duality

Although electromagnetic waves exhibit many characteristics and properties of waves, they can also exhibit some characteristics and properties of particles. We call these "particles" **photons**. Because of this, we say that light (and all EM radiation) has a dual nature. At times, light acts like a wave, and at other times it acts like a particle.

Characteristics of light that indicate light behaves like a wave include:

- Diffraction
- Interference
- Doppler Effect
- Young's Double-Slit Experiment

Characteristics of light that indicate light also acts as a particle include Blackbody Radiation, the Photoelectric Effect, and the Compton Effect.

12.01 Q: Light demonstrates the characteristics of

    (1) particles, only

    (2) waves, only

    (3) both particles and waves

    (4) neither particles nor waves

12.01 A: (3) both particles and waves.

12.02 Q: Which phenomenon provides evidence that light has a wave nature?

(1) emission of light from an energy-level transition in a hydrogen atom

(2) diffraction of light passing through a narrow opening

(3) absorption of light by a black sheet of paper

(4) reflection of light from a mirror

12.02 A: (2) diffraction is a phenomenon only applicable to waves.

Blackbody Radiation

The radiation emitted from a very hot object (known as black-body radiation) didn't align with physicists' understanding of light as a wave. Specifically, very hot objects emitted radiation in a specific spectrum of frequencies and intensities, which varied with the temperature of the object. Hotter objects had higher intensities at lower wavelengths (toward the blue/UV end of the spectrum), and cooler objects emitted more intensity at higher wavelengths (toward the red/infrared end of the spectrum). Physicists expected that at very short wavelengths the energy radiated would become very large, in contrast to observed spectra. This problem was known as the ultraviolet catastrophe.

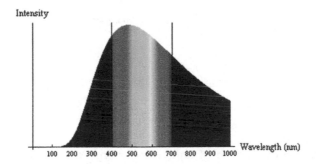

German physicist Max Planck solved this puzzle by proposing that atoms could only absorb or emit radiation In specific, non-continuous amounts, known as quanta. Energy, therefore, is quantized - it only exists in specific discrete amounts. For his work, Planck was awarded the Nobel Prize in Physics in 1918.

Photoelectric Effect

Further evidence that light behaves like a particle was proposed by Albert Einstein in 1905. Scientists had observed that when EM radiation struck a piece of metal, electrons could be emitted (known as **photoelectrons**). What was troubling was that not all EM radiation created photoelectrons. Regardless of what intensity of light was incident upon the metal, the only variable that effected the creation of photoelectrons was the frequency of the light.

If energy exists only in specific, discrete amounts, EM radiation exists in specific discrete amounts, and we call these smallest possible "pieces" of EM radiation "photons." A photon has zero mass and zero charge, and because it is a type of EM radiation, its velocity in a vacuum is equal to c (3×10^8 m/s). The energy of each photon of light is therefore quantized and is related to its frequency by the equation:

$$E_{photon} = hf = \frac{hc}{\lambda}$$

In this equation, the value of h, known as **Planck's Constant**, is given as 6.63×10^{-34} J•s and is available from the Regents Physics Reference Table.

Einstein proposed that the electrons in the metal object were held in an "energy well," and had to absorb at least enough energy to pull the electron out of the energy well in order to emit a photoelectron. The electrons in the metal would not be released unless they absorbed a single photon with that minimum amount of energy, known as the work function of the metal. Any excess absorbed energy beyond that required to free the electron became kinetic energy for the photoelectron.

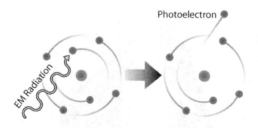

When a high-energy photon of light with energy greater than the energy holding an electron to its nucleus is absorbed by an atom, the electron is emitted as a photoelectron. The kinetic energy of the emitted photoelectron is exactly equal to the amount of energy holding the electron to the nucleus subtracted from the energy of the absorbed photon.

This theory extended Planck's work and inferred the particle-like behavior of photons of light. Photoelectrons would be ejected from the metal only if they absorbed a photon of light with frequency greater than or equal to a minimum threshold frequency, corresponding to the energy of a photon equal to the metal's "electron well" energy for the most loosely held electrons. Regardless of the intensity of the incident EM radiation, only EM radiation at or above the threshold frequency could produce photoelectrons.

12.03 Q: A photon of light traveling through space with a wavelength of 6×10^{-7} meters has an energy of

(1) 4.0×10^{-40} J

(2) 3.3×10^{-19} J

(3) 5.4×10^{10} J

(4) 5.0×10^{14} J

12.03 A: (2) $E = \dfrac{hc}{\lambda} = \dfrac{(6.63 \times 10^{-34}\, J \bullet s)(3 \times 10^{8}\, {}^{m}\!/\!_{s})}{6 \times 10^{-7}\, m} = 3.3 \times 10^{-19}\, J$

12.04 Q: The spectrum of visible light emitted during transitions in excited hydrogen atoms is composed of blue, green, red, and violet lines. What characteristic of light determines the amount of energy carried by a photon of that light?

(1) amplitude

(2) frequency

(3) phase

(4) velocity

12.04 A: (2) frequency determines the energy carried by a photon.

12.05 Q: Determine the frequency of a photon whose energy is 3×10^{-19} joule.

12.05 A: $E = hf$

$$f = \frac{E}{h} = \frac{3 \times 10^{-19}\, J}{6.63 \times 10^{-34}\, J \bullet s} - 4.5 \times 10^{14}\, Hz$$

12.06 Q: Light of wavelength 5.0×10^{-7} meter consists of photons having an energy of

(1) 1.1×10^{-48} J

(2) 1.3×10^{-27} J

(3) 4.0×10^{-19} J

(4) 1.7×10^{-5} J

12.06 A: (3) $E = \dfrac{hc}{\lambda} = \dfrac{(6.63 \times 10^{-34}\, J \bullet s)(3 \times 10^{8}\, {}^{m}\!/\!_{s})}{5 \times 10^{-7}\, m} = 4 \times 10^{-19}\, J$

12.07 Q: A photon has a wavelength of 9×10^{-10} meters. Calculate the energy of this photon in joules.

12.07 A: $E = \dfrac{hc}{\lambda} = \dfrac{(6.63 \times 10^{-34} J \bullet s)(3 \times 10^{8} \, ^m\!/_s)}{9 \times 10^{-10} \, m} = 2.2 \times 10^{-16} J$

12.08 Q: The graph below represents the relationship between the energy and the frequency of photons.

Energy vs. Frequency

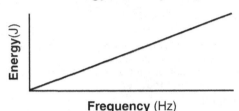

Frequency (Hz)

The slope of the graph would be

(1) 6.63×10^{-34} J•s

(2) 6.67×10^{-11} N•m2/kg2

(3) 1.60×10^{-19} J

(4) 1.60×10^{-19} C

12.08 A: (1) The slope of the graph, rise over run, is equivalent to the energy divided by the frequency, which gives you Planck's constant.

12.09 Q: The alpha line in the Balmer series of the hydrogen spectrum consists of light having a wavelength of 6.56×10^{-7} meter.

(A) Calculate the frequency of this light.

(B) Determine the energy in joules of a photon of this light.

(C) Determine the energy in electronvolts of a photon of this light.

12.09 A: (A) $v = f\lambda$

$$f = \frac{v}{\lambda} = \frac{3 \times 10^{8} \, ^m\!/_s}{6.56 \times 10^{-7} \, m} = 4.57 \times 10^{14} \, Hz$$

(B) $E = \dfrac{hc}{\lambda} = \dfrac{(6.63 \times 10^{-34} J \bullet s)(3 \times 10^{8} \, ^m\!/_s)}{6.56 \times 10^{-7} \, m} = 3.03 \times 10^{-19} J$

(C) $3.03 \times 10^{-19} J \times \dfrac{1eV}{1.6 \times 10^{-19} J} = 1.89eV$

de Broglie Wavelength

Einstein continued to extend his theories around the interaction of photons and atomic particles, going so far as to hypothesize that photons could have momentum, also a particle property, even though they had no mass.

In 1922, American physicist Arthur Compton shot an X-ray photon at a graphite target to observe the collision between the photon and one of the graphite atom's electrons. Compton observed that when the photon collided with an electron, a photoelectron was emitted, but the original X-ray was also scattered and emitted, and with a longer wavelength (indicating it had lost energy).

Further, the longer wavelength also indicated that the photon must have lost momentum. A detailed analysis showed that the energy and momentum lost by the X-ray was exactly equal to the energy and momentum gained by the photoelectron. Compton therefore concluded that not only do photons have momentum, they also obey the laws of conservation of energy and conservation of momentum!

In 1923, French physicist Louis de Broglie took Compton's finding one step further. He stated that if EM waves can behave as moving particles, it would only make sense that a moving particle should exhibit wave properties. De Broglie's hypothesis was confirmed by shooting electrons through a double slit, similar to Young's Double Slit Experiment, and observing a diffraction pattern. The smaller the particle, the more apparent its wave properties are. The wavelength of a moving particle, now known as the de Broglie Wavelength, is given by:

$$\lambda = \frac{h}{p}$$

12.10 Q: Moving electrons are found to exhibit properties of
    (1) particles, only
    (2) waves, only
    (3) both particles and waves
    (4) neither particles nor waves

12.10 A: (3) moving particles have both particle and wave properties.

12.11 Q: Which phenomenon best supports the theory that matter has a wave nature?
    (1) electron momentum
    (2) electron diffraction
    (3) photon momentum
    (4) photon diffraction

12.11 A: (2) The diffraction of electrons indicates that electrons behave like waves.

12.12 Q: Wave-particle duality is most apparent in analyzing the motion of
(1) a baseball
(2) a space shuttle
(3) a galaxy
(4) an electron

12.12 A: (4) Wave-particle duality is most easily observed in small objects.

Models of the Atom

In the early 1900s, scientists around the world began to refine and revise our understanding of atomic structure and sub-atomic particles. Scientists understood that matter was made up of atoms, and J.J. Thompson had shown that atoms contained very small negative particles known as electrons, but beyond that, the atom remained a mystery.

New Zealand scientist Ernest Rutherford devised an experiment to better understand the rest of the atom. The experiment, known as Rutherford's Gold Foil Experiment, involved shooting alpha particles (helium nuclei) at a very thin sheet of gold foil and observing the deflection of the particles after passing through the gold foil. Rutherford found that although most of the particles went through undeflected, a significant number of alpha particles were deflected by large amounts. Using an analysis based around Coulomb's Law and the conservation of momentum, Rutherford concluded that:

1. Atoms have a small, massive, positive nucleus at the center.
2. Electrons must orbit the nucleus.
3. Most of the atom is made up of empty space.

Rutherford's model was incomplete, though, in that it didn't account for a number of effects predicted by classical physics. Classical physics predicted that if the electron orbits the atom, it is constantly accelerating, and should therefore emit photons of EM radiation. Because the atom emits photons, it should be losing energy, therefore the orbit of the electron would quickly decay into the nucleus and the atom would be unstable. Further, elements were found to emit and absorb EM radiation only at specific frequencies, which did not correlate to Rutherford's theory.

Following Rutherford's discovery, Danish physicist Niels Bohr traveled to England to join Rutherford's research group and refine Rutherford's model of the atom. Instead of focusing on all atoms, Bohr confined his research to developing a model of the simple hydrogen atom. Bohr's model made the following assumptions:

1. Electrons don't lose energy as they accelerate around the nucleus. Instead, energy is quantized. Electrons can only exist at specific discrete energy levels.
2. Each atom allows only a limited number of specific orbits (electrons) at each energy level.
3. To change energy levels, an electron must absorb or emit a photon of energy exactly equal to the difference between the electron's initial and final energy levels: $E_{photon} = E_i - E_f$

Bohr's Model, therefore, was able to explain some of the limitations of Rutherford's Model. Further, Bohr was able to use his model to predict the frequencies of photons emitted and absorbed by hydrogen, explaining Rutherford's problem of emission and absorption spectra! For his work, Bohr was awarded the Nobel Prize in Physics in 1922.

12.13 Q: Calculate the energy and wavelength of the emitted photon when an electron moves from an energy level of -1.51 eV to -13.6 eV.

12.13 A: $E_{photon} = E_i - E_f = (-1.51eV) - (-13.6eV) = 12.09eV$

$$E_{photon} = \frac{hc}{\lambda}$$

$$\lambda = \frac{hc}{E_{photon}} = \frac{(6.63 \times 10^{-34} J \bullet s)(3 \times 10^8 \, ^m/_s)}{(12.09eV)(1.6 \times 10^{-19} \, ^J/_{eV})} = 1.03 \times 10^{-7} m$$

Energy Level Diagrams

A useful tool for visualizing the allowed energy levels in an atom is the energy level diagram. Two of these diagrams (one for hydrogen and one for mercury) are provided for you on your reference table. In each of these diagrams, the n=1 energy state is the lowest possible energy for an electron of that atom, known as the ground state. The energy corresponding to n=1 is shown on the right side of the diagram in electronvolts. So, for hydrogen, the ground state is a level of -13.6 eV.

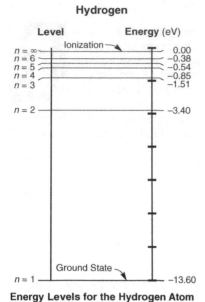

Hydrogen

Energy Levels for the Hydrogen Atom

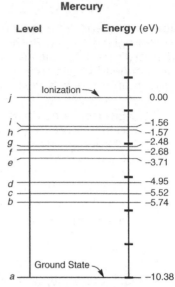

Mercury

A Few Energy Levels for the Mercury Atom

The energy levels are negative to indicate that the electron is bound by the nucleus of the atom. If the electron reaches 0 eV, it is no longer bound by the atom and can be emitted as a photoelectron (i.e. the atom becomes ionized). Any remaining energy becomes the kinetic energy of the photoelectron.

12.14 Q: An electron in a hydrogen atom drops from the n=3 to the n=2 state. Determine the energy of the emitted radiation.

12.14 A:

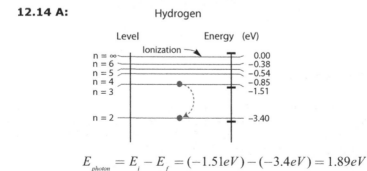

Hydrogen

$$E_{photon} = E_i - E_f = (-1.51eV) - (-3.4eV) = 1.89eV$$

12.15 Q: Which type of photon is emitted when an electron in a hydrogen atom drops from the n = 2 to the n = 1 energy level?

(1) ultraviolet

(2) visible light

(3) infrared

(4) radio wave

12.15 A: (1) First find the amount of energy emitted in electron volts, convert that energy to Joules, then find the frequency of the emitted radiation, which you can look up on the EM Spectrum to determine the radiation type.

$$E_{photon} = E_i - E_f = (-3.4eV) - (-13.6eV) = 10.2eV$$

$$10.2eV \times \frac{1.6 \times 10^{-19} J}{1eV} = 1.63 \times 10^{-18} J$$

$$E = hf \quad f = \frac{E}{h} = \frac{1.63 \times 10^{-18} J}{6.63 \times 10^{-34} J \bullet s} = 2.46 \times 10^{15} Hz$$

12.16 Q: Base your answers on the Energy Level Diagram for Hydrogen in the Regents Physics Reference Table.

(A) Determine the energy, in electronvolts, of a photon emitted by an electron as it moves from the n = 6 to the n = 2 energy level in a hydrogen atom.

(B) Convert the energy of the photon to joules.

(C) Calculate the frequency of the emitted photon.

(D) Is this the only energy and/or frequency that an electron in the n = 6 energy level of a hydrogen atom could emit? Explain your answer.

12.16 A: (A) $E_{photon} = E_i - E_f = (-0.38eV) - (-3.4eV) = 3.02eV$

(B) $3.02eV \times \dfrac{1.6 \times 10^{-19} J}{1eV} = 4.83 \times 10^{-19} J$

(C) $E = hf \rightarrow f = \dfrac{E}{h} = \dfrac{4.83 \times 10^{-19} J}{6.63 \times 10^{-34} J \bullet s} = 7.29 \times 10^{14} Hz$

(D) No, this is not the only energy and/or frequency that an electron in the n=6 energy level of a hydrogen atom could emit. The electron can return to any of the five lower energy levels.

12.17 Q: An electron in a mercury atom drops from energy level f to energy level c by emitting a photon having an energy of

(1) 8.20 eV

(2) 5.52 eV

(3) 2.84 eV

(4) 2.68 eV

12.17 A: (3) $E_{photon} = E_i - E_f = (-2.68eV) - (-5.52eV) = 2.84eV$

12.18 Q: A mercury atom in the ground state absorbs 20 electronvolts of energy and is ionized by losing an electron. How much kinetic energy does this electron have after the ionization?

(1) 6.40 eV

(2) 9.62 eV

(3) 10.38 eV

(4) 13.60 eV

12.18 A: (2) The ionization energy for an electron in the ground state of a mercury atom is 10.38 eV according to the Mercury Energy Level Diagram in the Regents Physics Reference Table. If the atom absorbs 20 eV of energy, and uses up 10.38 eV in ionizing the electron, the electron has a leftover energy of 9.62 eV, which must be the electron's kinetic energy.

12.19 Q: A hydrogen atom with an electron initially in the n = 2 level is excited further until the electron is in the n = 4 level. This energy level change occurs because the atom has

(1) absorbed a 0.85-eV photon

(2) emitted a 0.85-eV photon

(3) absorbed a 2.55-eV photon

(4) emitted a 2.55-eV photon

12.19 A: (3) absorbed a 2.55-eV photon.

Atomic Spectra

Once you understand the energy level diagram, it quickly becomes obvious that atoms can only emit certain frequencies of photons, correlating to the difference between energy levels as an electron falls from a higher energy state to a lower energy state. In similar fashion, electrons can only absorb photons with energy equal to the difference in energy levels as the electron jumps from a lower to a higher energy state. This leads to unique atomic spectra of emitted radiation for each element.

An object that is heated to the point where it glows (**incandescence**) emits a continuous energy spectrum, described as blackbody radiation.

If a gas-discharge lamp is made from mercury vapor, the mercury vapor is made to emit light by application of a high electrical potential. The light emitted by the mercury vapor is created by electrons in higher energy states falling to lower energy states, therefore the photons emitted correspond directly in wavelength to the difference in energy levels of the electrons. This creates a unique spectrum of frequencies which can be observed by separating the colors using a prism, known as an emission spectrum. By analyzing the emission spectra of various objects, scientists can determine the composition of those objects.

In similar fashion, if light of all colors is shone through a cold gas, the gas will only absorb the frequencies corresponding to photon energies exactly equal to the difference between the gas's atomic energy levels. This creates a spectrum with all colors except those absorbed by the gas, known as an absorption spectrum.

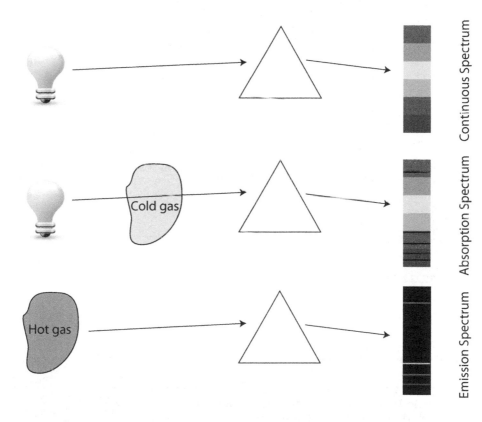

12.20 Q: The bright-line emission spectrum of an element can best be explained by

(1) electrons transitioning between discrete energy levels in the atoms of that element

(2) protons acting as both particles and waves

(3) electrons being located in the nucleus

(4) protons being dispersed uniformly throughout the atoms of that element

12.20 A: (1) bright-line emission spectra are created by electrons moving between energy levels, giving off photons of energy equal to the difference in energy levels.

12.21 Q: The diagram below represents the bright-line spectra of four elements, A, B, C, and D, and the spectrum of an unknown gaseous sample.

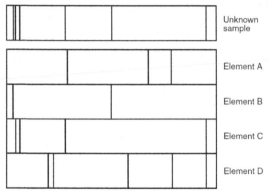

Based on comparisons of these spectra, which two elements are found in the unknown sample?

(1) A and B

(2) A and D

(3) B and C

(4) C and D

12.21 A: (3) Elements B and C have bright lines corresponding to the unknown sample.

Mass—Energy Equivalence

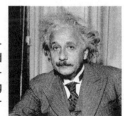

In 1905, in a paper titled "Does the Inertia of a Body Depend Upon Its Energy Content," Albert Einstein proposed the revolutionary concept that an object's mass is a measure of how much energy that object contains, opening a door to a host of world-changing developments, eventually leading us to the major understanding that the source of all energy in the universe is, ultimately, the conversion of mass into energy!

If mass is a measure of an object's energy, we need to re-evaluate our statements of the law of conservation of mass and the law of conservation of energy. Up to this point, we have thought of these as separate statements of fact in the universe. Based on Einstein's discovery, however, mass and energy are two concepts effectively describing the same thing, therefore we could more appropriately combine these two laws into a single law: the law of conservation of mass-energy. This law states that mass-energy cannot be created nor destroyed.

The concept of mass-energy is one that is often misunderstood and oftentimes argued in terms of semantics. For example, a popular argument states that the concept of mass-energy equivalence means that mass can be converted to energy, and energy can be converted to mass. Many would disagree that this can occur, countering that since mass and energy are effectively the same thing, you can't convert one to the other. For our purposes, we'll save these arguments for future courses of study. Instead, we will focus on a basic conceptual understanding.

The universal conservation laws we have studied so far this course include:

- Conservation of Mass-Energy
- Conservation of Charge
- Conservation of Momentum

Einstein's famous formula, E=mc², relates the amount of energy contained in matter to the mass times the speed of light in a vacuum (c=3×10⁸ m/s) squared. Theoretically, then, we could determine the amount of energy represented by 1 kilogram of matter as follows:

12.22 Q: What is the energy equivalent of 1 kilogram of matter?

12.22 A: $E = mc^2 = (1kg)(3 \times 10^8 \, m/s)^2 = 9 \times 10^{16} \, J$

This is a very large amount of energy. To put it in perspective, the energy equivalent of a large pickup truck is in the same order of magnitude of the total annual energy consumption of the United States!

More practically, however, it is not realistic to convert large quantities of mass completely into energy. Current practice revolves around converting small amounts of mass into energy in nuclear processes. Typically these masses are so small that measuring in units of kilograms is cumbersome. Instead, scientists often work with the much smaller **universal mass unit** (u), which is equal in mass to one-twelfth the mass of a single atom of Carbon-12. The mass of a proton and neutron, therefore, is close to 1u, and the mass of an electron is close to 5×10^{-4}u. In precise terms, $1u=1.66053886\times10^{-27}$kg.

One universal mass unit (1u) completely converted to energy is equivalent to 931 MeV. Because mass and energy are different forms of the same thing, this could even be considered a unit conversion problem. If given a mass in universal mass units, you can use this equivalence directly from the front of the Regents Physics Reference Table to solve for the equivalent amount of energy, without having to convert into standard units and utilize the $E=mc^2$ equation.

12.23 Q: If a deuterium nucleus has a mass of 1.53×10^{-3} universal mass units (u) less than its components, how much energy does its mass represent?

12.23 A: $(1.53\times10^{-3}u)\times\dfrac{9.31\times10^2\,MeV}{1u}=1.42\,MeV$

The nucleus of an atom consists of positively charged protons and neutral neutrons. Collectively, these nuclear particles are known as nucleons. Protons repel each other electrically, so why doesn't the nucleus fly apart? There is another force which holds nucleons together, known as the **strong nuclear force**. This extremely strong force overcomes the electrical repulsion of the protons, but it is only effective over very small distances.

Because nucleons are held together by the strong nuclear force, you must add energy to the system to break apart the nucleus. The energy required to break apart the nucleus is known as the **binding energy** of the nucleus.

If measured carefully, we find that the mass of a stable nucleus is actually slightly less than the mass of its individual component nucleons. The difference in mass between the entire nucleus and the sum of its component parts is known as the **mass defect** (Δm). The binding energy of the nucleus, therefore, must be the energy equivalent of the mass defect due to the law of conservation of mass-energy: $E_{binding}=\Delta mc^2$.

Fission is the process in which a nucleus splits into two or more nuclei. For heavy (larger) nuclei such as Uranium-235, the mass of the original nucleus is greater than the sum of the mass of the fission products. Where did this mass go? It is released as energy! A commonly used fission reaction involves shooting a neutron at an atom of Uranium-235, which briefly becomes Uranium-236, an unstable isotope. The Uranium-236 atom then fissions into a Barium-141 atom and a Krypton-92 atom, releasing its excess energy while also sending out three more neutrons to continue a chain reaction! This process is responsible for our nuclear power plants, and is also the basis (in an uncontrolled reaction) of atomic fission bombs.

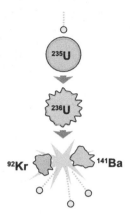

Fusion, on the other hand, is the process of combining two or more smaller nuclei into a larger nucleus. If this occurs with small nuclei, the product of the reaction may have a smaller mass its precursors, thereby releasing energy as part of the reaction. This is the basic nuclear reaction that fuels our sun and the stars as hydrogen atoms combine to form helium. This is also the basis of atomic hydrogen bombs.

Nuclear fusion holds tremendous potential as a clean source of power with widely available source material (we can create hydrogen from water). The most promising fusion reaction for controlled energy production fuses two isotopes of hydrogen known as deuterium and tritium to form a helium nucleus and a neutron, as well as an extra neutron, while releasing a considerable amount of energy. Currently, creating a sustainable, controlled fusion reaction that outputs more energy than is required to start the reaction has not yet been demonstrated, but remains an area of focus for scientists and engineers.

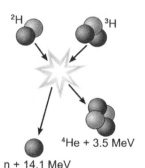

12.24 Q: In the first nuclear reaction using a particle accelerator, accelerated protons bombarded lithium atoms, producing alpha particles and energy. The energy resulted from the conversion of mass into energy. The reaction can be written as shown below.

$$_1^1H + {_3^7}Li \rightarrow {_2^4}He + energy$$

Data Table

Particle	Symbol	Mass (u)
proton	$_1^1H$	1.007 83
lithium atom	$_3^7Li$	7.016 00
alpha particle	$_2^4He$	4.002 60

(A) Determine the difference between the total mass of a proton plus a lithium atom, and the total mass of two alpha particles, in universal mass units.

(B) Determine the energy in megaelectronvolts produced in the reaction of a proton with a lithium atom.

12.24 A: (A) $(1.00783u + 7.01600u) - 2(4.00260u) = 0.01863u$

(B) $0.01863u \times \dfrac{9.31 \times 10^2 \; MeV}{1u} = 17.3 \, MeV$

12.25 Q: The energy produced by the complete conversion of 2×10^{-5} kilograms of mass into energy is

(1) 1.8 TJ

(2) 6.0 GJ

(3) 1.8 MJ

(4) 6.0 kJ

12.25 A: (1) $E = mc^2 = (2 \times 10^{-5} kg)(3 \times 10^8 \; ^m\!/_s)^2 = 1.8 \times 10^{12} J = 1.8 TJ$

12.26 Q: A tritium nucleus is formed by combining two neutrons and a proton. The mass of this nucleus is 9.106×10^{-3} universal mass unit less than the combined mass of the particles from which it is formed. Approximately how much energy is released when this nucleus is formed?

(1) 8.48 × 10⁻² MeV

(2) 2.73 MeV

(3) 8.48 MeV

(4) 273 MeV

12.26 A: (3) $9.106 \times 10^{-3} u \times \dfrac{9.31 \times 10^2 \; MeV}{1u} = 8.48 \, MeV$

12.27 Q: The energy equivalent of 5×10^{-3} kilogram is

(1) 8.0×10⁵J

(2) 1.5×10⁶J

(3) 4.5×10¹⁴ J

(4) 3.0×10¹⁹ J

12.27 A: (3) $E = mc^2 = (5 \times 10^{-3} kg)(3 \times 10^8 \; ^m\!/_s)^2 = 4.5 \times 10^{14} J$

12.28 Q: After a uranium nucleus emits an alpha particle, the total mass of the new nucleus and the alpha particle is less than the mass of the original uranium nucleus. Explain what happens to the missing mass.

12.28 A: The missing mass is converted into energy.

The Standard Model

As we've learned previously, the atom is the smallest part of an element (such as oxygen) that has the characteristics of the element. Atoms are made up of very small negatively charged electrons surrounding the much larger nucleus. The nucleus is composed of positively charged protons and neutral neutrons. The positively charged protons exert a repelling electrical force upon each other, but the strong nuclear force holds the protons and neutrons together in the nucleus.

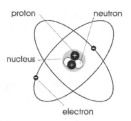

This completely summarized our understanding of atomic structure until the 1930s, when scientists began to discover evidence that there was more to the picture and that protons and nucleons were made up of even smaller particles. This launched the particle physics movement, which, to this day, continues to challenge our understanding of the entire universe by exploring the structure of the atom.

In addition to matter we're familiar with, researchers have discovered the existence of antimatter. **Antimatter** is matter made up of particles with the same mass as regular matter particles, but opposite charges and other characteristics. An **antiproton** is a particle with the same mass as a proton, but a negative (opposite) charge. A **positron** has the same mass as an electron, but a positive charge. An **antineutron** has the same mass as a neutron, but has other characteristics opposite that of the neutron.

When a matter particle and its corresponding antimatter particle meet, the particles may combine to **annihilate** each other, resulting in the complete conversion of both particles into energy consistent with the mass-energy equivalence equation: E=mc².

12.29 Q: A proton and an antiproton collide and completely annihilate each other. How much energy is released? ($m_{proton}=1.67\times10^{-27}$kg)

12.29 A: $E = mc^2 = 2(1.67\times10^{-27}kg)(3\times10^8\,{}^{m}\!/_{s})^2 = 3\times10^{-10}J$

We've dealt with many types of forces in this course, ranging from contact forces such as tensions and normal forces to field forces such as the electrical force and gravitational force. When observed from their most basic aspects, however, we can consolidate all observed forces in the universe into the following four known fundamental forces. They are, from strongest to weakest:

1. Strong Nuclear Force: holds protons and neutrons together in the nucleus
2. Electromagnetic Force: electrical and magnetic attraction and repulsion
3. Weak force: responsible for radioactive beta decay
4. Gravitational Force: attractive force between objects with mass

Understanding these forces remains a topic of scientific research, with current work exploring the possibility that forces are actually conveyed by an exchange of force-carrying particles such as photons, bosons, gluons, and gravitons.

12.30 Q: The particles in a nucleus are held together primarily by the
(1) strong force
(2) gravitational force
(3) electrostatic force
(4) magnetic force

12.30 A: (1) the strong nuclear force holds protons and neutrons together in the nucleus.

12.31 Q: Which fundamental force is primarily responsible for the attraction between protons and electrons?
(1) strong
(2) weak
(3) gravitational
(4) electromagnetic

12.31 A: (4) the electromagnetic force is responsible for the electrostatic attraction and repulsion of charged particles.

12.32 Q: Which statement is true of the strong nuclear force?
(1) It acts over very great distances.
(2) It holds protons and neutrons together.
(3) It is much weaker than gravitational forces.
(4) It repels neutral charges.

12.32 A: (2) The strong nuclear force holds protons and neutrons together.

12.33 Q: The strong force is the force of

    (1) repulsion between protons

    (2) attraction between protons and electrons

    (3) repulsion between nucleons

    (4) attraction between nucleons

12.33 A: (4) attraction between nucleons (nucleons are particles in the nucleus such as protons and neutrons).

The current model of sub-atomic structure used to understand matter is known as the Standard Model. Development of this model began in the late 1960s, and has continued through today with contributions from many scientists across the world. The Standard Model explains the interactions of the strong (nuclear), electromagnetic, and weak forces, but has yet to account for the gravitational force. The search for the theorized Higgs Boson at Fermilab and CERN is an attempt to better unify and strengthen the Standard Model.

Although the Standard Model itself is a very complicated theory, the basic structure of the model is fairly straightforward. According to the model, all matter is divided into two categories, known as **hadrons** and the much smaller **leptons**. All of the fundamental forces act on hadrons, which include particles such as protons and neutrons. In contrast, the strong nuclear force doesn't act on leptons, so only three fundamental forces act on leptons such as electrons, positrons, muons, tau particles and neutrinos.

Hadrons are further divided into **baryons** and **mesons**. Baryons such as protons and neutrons are composed of three smaller particles known as **quarks**. Charges of baryons are always whole numbers. Mesons are composed of a quark and an anti-quark (for example, an up quark and an anti-down quark). If this sounds like a lot to keep track of, have no fear, this is summarized for you on the Regents Physics Reference Table.

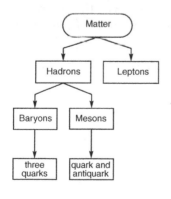

Classification of Matter

Particles of the Standard Model

Quarks

Name	up	charm	top
Symbol	u	c	t
Charge	$+\frac{2}{3}e$	$+\frac{2}{3}e$	$+\frac{2}{3}e$

down	strange	bottom
d	s	b
$-\frac{1}{3}e$	$-\frac{1}{3}e$	$-\frac{1}{3}e$

Leptons

electron	muon	tau
e	μ	τ
$-1e$	$-1e$	$-1e$

electron neutrino	muon neutrino	tau neutrino
ν_e	ν_μ	ν_τ
0	0	0

Note: For each particle, there is a corresponding antiparticle with a charge opposite that of its associated particle.

Scientists have identified six types of quarks. For each of the six types of quarks, there also exists a corresponding anti-quark with an opposite charge. The quarks have rather interesting names: up quark, down quark, charm quark, strange quark, top quark, and bottom quark. Charges on each quark are either one third of an elementary charge, or two thirds of an elementary charge, positive or negative, and the quarks are symbolized by the first letter of their name. For the associated anti-quark, the symbol is the first letter of the anti-quark's name, with a line over the name. For example, the symbol for the up quark is u. The symbol for the anti-up quark is ū.

Similarly, scientists have identified six types of leptons: the electron, the muon, the tau particle, and the electron neutrino, muon neutrino, and tau neutrino. Again, for each of these leptons there also exists an associated anti-lepton. The most familiar lepton, the electron, has a charge of -1e. Its anti-particle, the positron, has a charge of +1e.

Since a proton is made up of three quarks, and has a positive charge, the sum of the charges on its constituent quarks must be equal to one elementary charge. A proton is actually comprised of two up quarks and one down quark. We can verify this by adding up the charges of the proton's constituent quarks (uud).

$$\left(+\frac{2}{3}e\right) + \left(+\frac{2}{3}e\right) + \left(-\frac{1}{3}e\right) = +1e$$

12.34 Q: A neutron is composed of up and down quarks. How many of each type of quark are needed to make a neutron?

12.34 A: The charge on the neutron must sum to zero, and the neutron is a baryon, so it is made up of three quarks. To achieve a total charge of zero, the neutron must be made up of one up quark (+2/3e) and two down quarks (-1/3e).

If the charge on a quark (such as the up quark) is (+2/3)e, the charge of the anti-quark (ū) is (-2/3)e. The anti-quark is the same type of particle, with the same mass, but with the opposite charge.

12.35 Q: What is the charge of the down anti-quark?

12.35 A: The down quark's charge is -1/3e, so the anti-down quark's charge must be +1/3e.

12.36 Q: Compared to the mass and charge of a proton, an antiproton has

(1) the same mass and the same charge

(2) greater mass and the same charge

(3) the same mass and the opposite charge

(4) greater mass and the opposite charge

12.36 A: (3) the same mass and the opposite charge.

12.37 Q: The diagram below represents the sequence of events (steps 1 through 10) resulting in the production of a D⁻ meson and a D⁺ meson. An electron and a positron (antielectron) collide (step 1), annihilate each other (step 2), and become energy (step 3). This energy produces an anticharm quark and a charm quark (step 4), which then split apart (steps 5 through 7). As they split, a down quark and an antidown quark are formed, leading to the final production of a D⁻ meson and a D⁺ meson (steps 8 through 10).

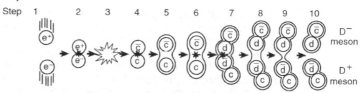

Adapted from: Electon/Positron Annihilation http:/www.particleadventure.org/frameless/eedd.html 7/23/2007

Which statement best describes the changes that occur in this sequence of events?

(1) Energy is converted into matter and then matter is converted into energy.

(2) Matter is converted into energy and then energy is converted into matter.

(3) Isolated quarks are being formed from baryons.

(4) Hadrons are being converted into leptons.

12.37 A: (2) Particles are converted into energy, which is then converted into particles.

12.38 Q: What fundamental force holds quarks together to form particles such as protons and neutrons?

(1) electromagnetic force

(2) gravitational force

(3) strong force

(4) weak force

12.38 A: (3) the strong force holds particles together in the nucleus.

12.39 Q: A particle unaffected by an electric field could have a quark composition of

(1) css

(2) bbb

(3) udc

(4) uud

12.39 A: (1) In order to not be affected by an electric field, the particle must be neutral. The only combination of quarks which results in a net charge of zero is css.

12.40 Q: For years, theoretical physicists have been refining a mathematical method called lattice quantum chromodynamics to enable them to predict the masses of particles consisting of various combinations of quarks and antiquarks. They recently used the theory to calculate the mass of the rare B_c particle, consisting of a charm quark and a bottom antiquark. The predicted mass of the B_c particle was about six times the mass of a proton.

Shortly after the prediction was made, physicists working at the Fermi National Accelerator Laboratory, Fermilab, were able to measure the mass of the B_c particle experimentally and found it to agree with the theoretical prediction to within a few tenths of a percent. In the experiment, the physicists sent beams of protons and antiprotons moving at 99.999% the speed of light in opposite directions around a ring 1.0 kilometer in radius. The protons and antiprotons were kept in their circular paths by powerful electromagnets. When the protons and antiprotons collided, their energy produced numerous new particles, including the elusive B_c.

These results indicate that lattice quantum chromodynamics is a powerful tool not only for confirming the masses of existing particles, but also for predicting the masses of particles that have yet to be discovered in the laboratory.

(A) Identify the class of matter to which the B_c particle belongs.

(B) Determine both the sign and the magnitude of the charge of the B_c particle in elementary charges.

(C) Explain how it is possible for a colliding proton and antiproton to produce a particle with six times the mass of either.

12.40 A: (A) Meson or Hadron

(B) +1e. The charge on a charm quark (+2/3e) and a bottom antiquark (+1/3e) sum to +1e.

(C) Energy is converted to mass.

12.41 Q: The charge of an antistrange quark is approximately
  (1) $+5.33 \times 10^{-20}$ C
  (2) -5.33×10^{-20} C
  (3) $+5.33 \times 10^{20}$ C
  (4) -5.33×10^{20} C

12.41 A: (1) charge on an antistrange quark is $+1/3e = +5.33 \times 10^{-20}$ C.

You can find more practice problems on the APlusPhysics website at: http://www.aplusphysics.com/regents.

Appendix A: Reference Table

THE UNIVERSITY OF THE STATE OF NEW YORK • THE STATE EDUCATION DEPARTMENT • ALBANY, NY 12234

Reference Tables for Physical Setting/PHYSICS
2006 Edition

List of Physical Constants		
Name	**Symbol**	**Value**
Universal gravitational constant	G	6.67×10^{-11} N•m^2/kg^2
Acceleration due to gravity	g	9.81 m/s^2
Speed of light in a vacuum	c	3.00×10^8 m/s
Speed of sound in air at STP		3.31×10^2 m/s
Mass of Earth		5.98×10^{24} kg
Mass of the Moon		7.35×10^{22} kg
Mean radius of Earth		6.37×10^6 m
Mean radius of the Moon		1.74×10^6 m
Mean distance—Earth to the Moon		3.84×10^8 m
Mean distance—Earth to the Sun		1.50×10^{11} m
Electrostatic constant	k	8.99×10^9 N•m^2/C^2
1 elementary charge	e	1.60×10^{-19} C
1 coulomb (C)		6.25×10^{18} elementary charges
1 electronvolt (eV)		1.60×10^{-19} J
Planck's constant	h	6.63×10^{-34} J•s
1 universal mass unit (u)		9.31×10^2 MeV
Rest mass of the electron	m_e	9.11×10^{-31} kg
Rest mass of the proton	m_p	1.67×10^{-27} kg
Rest mass of the neutron	m_n	1.67×10^{-27} kg

Prefixes for Powers of 10		
Prefix	**Symbol**	**Notation**
tera	T	10^{12}
giga	G	10^9
mega	M	10^6
kilo	k	10^3
deci	d	10^{-1}
centi	c	10^{-2}
milli	m	10^{-3}
micro	μ	10^{-6}
nano	n	10^{-9}
pico	p	10^{-12}

Approximate Coefficients of Friction		
	Kinetic	**Static**
Rubber on concrete (dry)	0.68	0.90
Rubber on concrete (wet)	0.58	
Rubber on asphalt (dry)	0.67	0.85
Rubber on asphalt (wet)	0.53	
Rubber on ice	0.15	
Waxed ski on snow	0.05	0.14
Wood on wood	0.30	0.42
Steel on steel	0.57	0.74
Copper on steel	0.36	0.53
Teflon on Teflon	0.04	

Appendix A: Reference Table

The Electromagnetic Spectrum

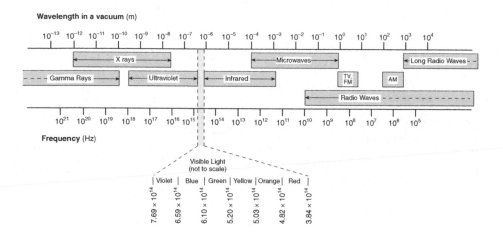

Absolute Indices of Refraction	
($f = 5.09 \times 10^{14}$ Hz)	
Air	1.00
Corn oil	1.47
Diamond	2.42
Ethyl alcohol	1.36
Glass, crown	1.52
Glass, flint	1.66
Glycerol	1.47
Lucite	1.50
Quartz, fused	1.46
Sodium chloride	1.54
Water	1.33
Zircon	1.92

Appendix A: Reference Table

Energy Level Diagrams

Hydrogen

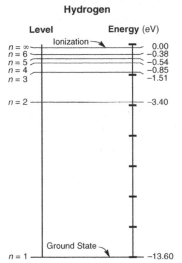

Energy Levels for the Hydrogen Atom

Level	Energy (eV)
$n = \infty$ Ionization	0.00
$n = 6$	−0.38
$n = 5$	−0.54
$n = 4$	−0.85
$n = 3$	−1.51
$n = 2$	−3.40
$n = 1$ Ground State	−13.60

Mercury

A Few Energy Levels for the Mercury Atom

Level	Energy (eV)
j Ionization	0.00
i	−1.56
h	−1.57
g	−2.48
f	−2.68
e	−3.71
d	−4.95
c	−5.52
b	−5.74
a Ground State	−10.38

Classification of Matter

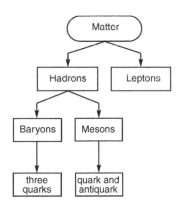

Matter → Hadrons, Leptons
Hadrons → Baryons, Mesons
Baryons → three quarks
Mesons → quark and antiquark

Particles of the Standard Model

Quarks

Name	up	charm	top
Symbol	u	c	t
Charge	$+\frac{2}{3}e$	$+\frac{2}{3}e$	$+\frac{2}{3}e$

down	strange	bottom
d	s	b
$-\frac{1}{3}e$	$-\frac{1}{3}e$	$-\frac{1}{3}e$

Leptons

electron	muon	tau
e	μ	τ
$-1e$	$-1e$	$-1e$

electron neutrino	muon neutrino	tau neutrino
ν_e	ν_μ	ν_τ
0	0	0

Note: For each particle, there is a corresponding antiparticle with a charge opposite that of its associated particle.

Appendix A: Reference Table

Electricity

$$F_e = \frac{kq_1q_2}{r^2}$$

$$E = \frac{F_e}{q}$$

$$V = \frac{W}{q}$$

$$I = \frac{\Delta q}{t}$$

$$R = \frac{V}{I}$$

$$R = \frac{\rho L}{A}$$

$$P = VI = I^2R = \frac{V^2}{R}$$

$$W = Pt = VIt = I^2Rt = \frac{V^2t}{R}$$

A = cross-sectional area
E = electric field strength
F_e = electrostatic force
I = current
k = electrostatic constant
L = length of conductor
P = electrical power
q = charge
R = resistance
R_{eq} = equivalent resistance
r = distance between centers
t = time
V = potential difference
W = work (electrical energy)
Δ = change
ρ = resistivity

Series Circuits

$$I = I_1 = I_2 = I_3 = \ldots$$

$$V = V_1 + V_2 + V_3 + \ldots$$

$$R_{eq} = R_1 + R_2 + R_3 + \ldots$$

Parallel Circuits

$$I = I_1 + I_2 + I_3 + \ldots$$

$$V = V_1 = V_2 = V_3 = \ldots$$

$$\frac{1}{R_{eq}} = \frac{1}{R_1} + \frac{1}{R_2} + \frac{1}{R_3} + \ldots$$

Circuit Symbols

—⊥— cell

≡ battery

——⁄— switch

—(V)— voltmeter

—(A)— ammeter

⌁⌁⌁ resistor

⌁⌁⌁ variable resistor

—(lll)— lamp

Resistivities at 20°C	
Material	**Resistivity ($\Omega \cdot m$)**
Aluminum	2.82×10^{-8}
Copper	1.72×10^{-8}
Gold	2.44×10^{-8}
Nichrome	$150. \times 10^{-8}$
Silver	1.59×10^{-8}
Tungsten	5.60×10^{-8}

Appendix A: Reference Table

Waves

$$v = f\lambda$$

$$T = \frac{1}{f}$$

$$\theta_i = \theta_r$$

$$n = \frac{c}{v}$$

$$n_1 \sin \theta_1 = n_2 \sin \theta_2$$

$$\frac{n_2}{n_1} = \frac{v_1}{v_2} = \frac{\lambda_1}{\lambda_2}$$

c = speed of light in a vacuum
f = frequency
n = absolute index of refraction
T = period
v = velocity or speed
λ = wavelength
θ = angle
θ_i = angle of incidence
θ_r = angle of reflection

Modern Physics

$$E_{photon} = hf = \frac{hc}{\lambda}$$

$$E_{photon} = E_i - E_f$$

$$E = mc^2$$

c = speed of light in a vacuum
E = energy
f = frequency
h = Planck's constant
m = mass
λ = wavelength

Geometry and Trigonometry

Rectangle

$$A = bh$$

Triangle

$$A = \frac{1}{2}bh$$

Circle

$$A = \pi r^2$$

$$C = 2\pi r$$

A = area
b = base
C = circumference
h = height
r = radius

Right Triangle

$$c^2 = a^2 + b^2$$

$$\sin \theta = \frac{a}{c}$$

$$\cos \theta = \frac{b}{c}$$

$$\tan \theta = \frac{a}{b}$$

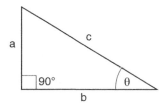

Appendix A: Reference Table

Mechanics

$$\bar{v} = \frac{d}{t}$$

$$a = \frac{\Delta v}{t}$$

$$v_f = v_i + at$$

$$d = v_i t + \frac{1}{2}at^2$$

$$v_f^2 = v_i^2 + 2ad$$

$$A_y = A \sin \theta$$

$$A_x = A \cos \theta$$

$$a = \frac{F_{net}}{m}$$

$$F_f = \mu F_N$$

$$F_g = \frac{Gm_1 m_2}{r^2}$$

$$g = \frac{F_g}{m}$$

$$p = mv$$

$$p_{before} = p_{after}$$

$$J = F_{net}t = \Delta p$$

$$F_s = kx$$

$$PE_s = \frac{1}{2}kx^2$$

$$F_c = ma_c$$

$$a_c = \frac{v^2}{r}$$

$$\Delta PE = mg\Delta h$$

$$KE = \frac{1}{2}mv^2$$

$$W = Fd = \Delta E_T$$

$$E_T = PE + KE + Q$$

$$P = \frac{W}{t} = \frac{Fd}{t} = F\bar{v}$$

a = acceleration

a_c = centripetal acceleration

A = any vector quantity

d = displacement or distance

E_T = total energy

F = force

F_c = centripetal force

F_f = force of friction

F_g = weight or force due to gravity

F_N = normal force

F_{net} = net force

F_s = force on a spring

g = acceleration due to gravity or gravitational field strength

G = universal gravitational constant

h = height

J = impulse

k = spring constant

KE = kinetic energy

m = mass

p = momentum

P = power

PE = potential energy

PE_s = potential energy stored in a spring

Q = internal energy

r = radius or distance between centers

t = time interval

v = velocity or speed

$\bar{v}$ = average velocity or average speed

W = work

x = change in spring length from the equilibrium position

Δ = change

θ = angle

μ = coefficient of friction

Appendix B: 2010 Regents Exam

Directions (1–35): For *each* statement or question, write in your answer booklet the *number* of the word or expression that, of those given, best completes the statement or answers the question.

1 A baseball player runs 27.4 meters from the batter's box to first base, overruns first base by 3.0 meters, and then returns to first base. Compared to the total distance traveled by the player, the magnitude of the player's total displacement from the batter's box is

(1) 3.0 m shorter   (3) 3.0 m longer
(2) 6.0 m shorter   (4) 6.0 m longer

2 A motorboat, which has a speed of 5.0 meters per second in still water, is headed east as it crosses a river flowing south at 3.3 meters per second. What is the magnitude of the boat's resultant velocity with respect to the starting point?

(1) 3.3 m/s   (3) 6.0 m/s
(2) 5.0 m/s   (4) 8.3 m/s

3 A car traveling on a straight road at 15.0 meters per second accelerates uniformly to a speed of 21.0 meters per second in 12.0 seconds. The total distance traveled by the car in this 12.0-second time interval is

(1) 36.0 m   (3) 216 m
(2) 180. m   (4) 252 m

4 A 0.149-kilogram baseball, initially moving at 15 meters per second, is brought to rest in 0.040 second by a baseball glove on a catcher's hand. The magnitude of the average force exerted on the ball by the glove is

(1) 2.2 N   (3) 17 N
(2) 2.9 N   (4) 56 N

5 Which body is in equilibrium?

(1) a satellite moving around Earth in a circular orbit
(2) a cart rolling down a frictionless incline
(3) an apple falling freely toward the surface of Earth
(4) a block sliding at constant velocity across a tabletop

6 As shown in the diagram below, a student standing on the roof of a 50.0-meter-high building kicks a stone at a horizontal speed of 4.00 meters per second.

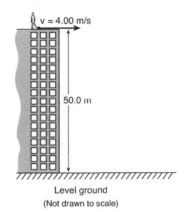

v = 4.00 m/s

50.0 m

Level ground
(Not drawn to scale)

How much time is required for the stone to reach the level ground below? [Neglect friction.]

(1) 3.19 s   (3) 10.2 s
(2) 5.10 s   (4) 12.5 s

7 On the surface of Earth, a spacecraft has a mass of 2.00×10^4 kilograms. What is the mass of the spacecraft at a distance of one Earth radius above Earth's surface?

(1) 5.00×10^3 kg   (3) 4.90×10^4 kg
(2) 2.00×10^4 kg   (4) 1.96×10^5 kg

8 A student pulls a 60.-newton sled with a force having a magnitude of 20. newtons. What is the magnitude of the force that the sled exerts on the student?

(1) 20. N   (3) 60. N
(2) 40. N   (4) 80. N

Appendix B: 2010 Regents Exam

9 The data table below lists the mass and speed of four different objects.

Data Table

Object	Mass (kg)	Speed (m/s)
A	4.0	6.0
B	6.0	5.0
C	8.0	3.0
D	16.0	1.5

Which object has the greatest inertia?
(1) A   (3) C
(2) B   (4) D

10 The diagram below shows a horizontal 12-newton force being applied to two blocks, A and B, initially at rest on a horizontal, frictionless surface. Block A has a mass of 1.0 kilogram and block B has a mass of 2.0 kilograms.

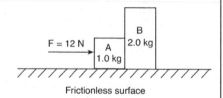

Frictionless surface

The magnitude of the acceleration of block B is
(1) 6.0 m/s^2 (3) 3.0 m/s^2
(2) 2.0 m/s^2 (4) 4.0 m/s^2

11 A ball is thrown vertically upward with an initial velocity of 29.4 meters per second. What is the maximum height reached by the ball? [Neglect friction.]

(1) 14.7 m   (3) 44.1 m
(2) 29.4 m   (4) 88.1 m

12 The diagram below represents a mass, m, being swung clockwise at constant speed in a horizontal circle.

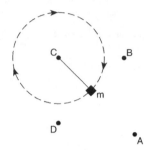

At the instant shown, the centripetal force acting on mass m is directed toward point
(1) A   (3) C
(2) B   (4) D

13 A 3.1-kilogram gun initially at rest is free to move. When a 0.015-kilogram bullet leaves the gun with a speed of 500. meters per second, what is the speed of the gun?

(1) 0.0 m/s   (3) 7.5 m/s
(2) 2.4 m/s   (4) 500. m/s

Appendix B: 2010 Regents Exam

14 Four projectiles, A, B, C, and D, were launched from, and returned to, level ground. The data table below shows the initial horizontal speed, initial vertical speed, and time of flight for each projectile.

Data Table

Projectile	Initial Horizontal Speed (m/s)	Initial Vertical Speed (m/s)	Time of Flight (s)
A	40.0	29.4	6.00
B	60.0	19.6	4.00
C	50.0	24.5	5.00
D	80.0	19.6	4.00

Which projectile traveled the greatest horizontal distance? [Neglect friction.]

(1) A   (3) C
(2) B   (4) D

15 A wound spring provides the energy to propel a toy car across a level floor. At time t_i, the car is moving at speed v_i across the floor and the spring is unwinding, as shown below. At time t_f, the spring has fully unwound and the car has coasted to a stop.

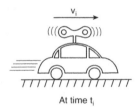

At time t_i At time t_f

Which statement best describes the transformation of energy that occurs between times t_i and t_f?

(1) Gravitational potential energy at t_i is converted to internal energy at t_f.

(2) Elastic potential energy at t_i is converted to kinetic energy at t_f.

(3) Both elastic potential energy and kinetic energy at t_i are converted to internal energy at t_f.

(4) Both kinetic energy and internal energy at t_i are converted to elastic potential energy at t_f.

16 A 75-kilogram bicyclist coasts down a hill at a constant speed of 12 meters per second. What is the kinetic energy of the bicyclist?

(1) 4.5×10^2 J   (3) 5.4×10^3 J
(2) 9.0×10^2 J   (4) 1.1×10^4 J

17 The diagram below represents a 155-newton box on a ramp. Applied force F causes the box to slide from point A to point B.

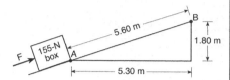

What is the total amount of gravitational potential energy gained by the box?

(1) 28.4 J   (3) 868 J
(2) 279 J   (4) 2740 J

18 An electric heater operating at 120. volts draws 8.00 amperes of current through its 15.0 ohms of resistance. The total amount of heat energy produced by the heater in 60.0 seconds is

(1) 7.20×10^3 J   (3) 8.64×10^4 J
(2) 5.76×10^4 J   (4) 6.91×10^6 J

19 Magnetic fields are produced by particles that are

(1) moving and charged
(2) moving and neutral
(3) stationary and charged
(4) stationary and neutral

20 A charge of 30. coulombs passes through a 24-ohm resistor in 6.0 seconds. What is the current through the resistor?

(1) 1.3 A   (3) 7.5 A
(2) 5.0 A   (4) 4.0 A

21 What is the magnitude of the electrostatic force between two electrons separated by a distance of 1.00×10^{-8} meter?

(1) 2.56×10^{-22} N   (3) 2.30×10^{-12} N
(2) 2.30×10^{-20} N   (4) 1.44×10^{-1} N

22 The diagram below represents the electric field surrounding two charged spheres, A and B.

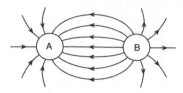

What is the sign of the charge of each sphere?

(1) Sphere A is positive and sphere B is negative.
(2) Sphere A is negative and sphere B is positive.
(3) Both spheres are positive.
(4) Both spheres are negative.

23 Which circuit has the *smallest* equivalent resistance?

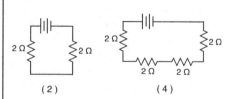

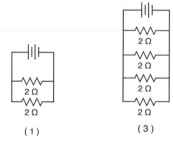

Appendix B: 2010 Regents Exam

Base your answers to questions 24 through 26 on the information and diagram below.

A longitudinal wave moves to the right through a uniform medium, as shown below. Points A, B, C, D, and E represent the positions of particles of the medium.

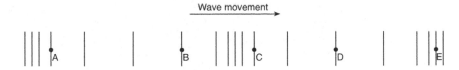

24 Which diagram best represents the motion of the particle at position C as the wave moves to the right?

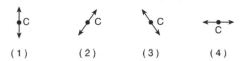

(1)   (2)   (3)   (4)

25 The wavelength of this wave is equal to the distance between points
(1) A and B (3) B and C
(2) A and C (4) B and E

26 The energy of this wave is related to its
(1) amplitude   (3) speed
(2) period   (4) wavelength

27 A ray of monochromatic yellow light ($f = 5.09 \times 10^{14}$ Hz) passes from water through flint glass and into medium X, as shown below.

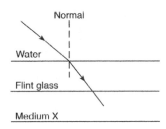

The absolute index of refraction of medium X is
(1) less than 1.33   (3) greater than 1.52 and less than 1.66
(2) greater than 1.33 and less than 1.52   (4) equal to 1.66

28 A light ray traveling in air enters a second medium and its speed slows to 1.71×10^8 meters per second. What is the absolute index of refraction of the second medium?

(1) 1.00　　　　　　　(3) 1.75
(2) 0.570　　　　　　(4) 1.94

29 Playing a certain musical note on a trumpet causes the spring on the bottom of a nearby snare drum to vibrate. This phenomenon is an example of

(1) resonance　　　　(3) reflection
(2) refraction　　　　(4) diffraction

30 In a vacuum, all electromagnetic waves have the same

(1) speed　　　　　　(3) frequency
(2) phase　　　　　　(4) wavelength

31 A particle that is composed of two up quarks and one down quark is a

(1) meson　　　　　　(3) proton
(2) neutron　　　　　(4) positron

32 A helium atom consists of two protons, two electrons, and two neutrons. In the helium atom, the strong force is a fundamental interaction between the

(1) electrons, only
(2) electrons and protons
(3) neutrons and electrons
(4) neutrons and protons

33 What total mass must be converted into energy to produce a gamma photon with an energy of 1.03×10^{-13} joule?

(1) 1.14×10^{-30}　　　(3) 3.09×10^{-5}
(2) 3.43×10^{-22}　　　(4) 8.75×10^{29}

34 Compared to the mass and charge of a proton, an antiproton has

(1) the same mass and the same charge
(2) greater mass and the same charge
(3) the same mass and the opposite charge
(4) greater mass and the opposite charge

Note that question 35 has only three choices.

35 As viewed from Earth, the light from a star has lower frequencies than the light emitted by the star because the star is

(1) moving toward Earth
(2) moving away from Earth
(3) stationary

Appendix B: 2010 Regents Exam

Answer all questions in this part.

Directions (36–50): For *each* statement or question, write in your answer booklet the *number* of the word or expression that, of those given, best completes the statement or answers the question.

36 The total work done in lifting a typical high school physics textbook a vertical distance of 0.10 meter is approximately

(1) 0.15 J   (3) 15 J
(2) 1.5 J   (4) 150 J

37 Which electrical unit is equivalent to one joule?

(1) volt per meter   (3) volt per coulomb
(2) ampere•volt   (4) coulomb•volt

38 A small electric motor is used to lift a 0.50-kilogram mass at constant speed. If the mass is lifted a vertical distance of 1.5 meters in 5.0 seconds, the average power developed by the motor is

(1) 0.15 W   (3) 3.8 W
(2) 1.5 W   (4) 7.5 W

39 A ball is dropped from the top of a cliff. Which graph best represents the relationship between the ball's total energy and elapsed time as the ball falls to the ground? [Neglect friction.]

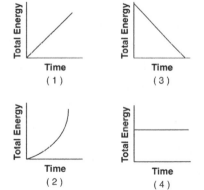

40 A child, starting from rest at the top of a playground slide, reaches a speed of 7.0 meters per second at the bottom of the slide. What is the vertical height of the slide? [Neglect friction.]

(1) 0.71 m   (3) 2.5 m
(2) 1.4 m   (4) 3.5 m

41 The graph below represents the relationship between the current in a metallic conductor and the potential difference across the conductor at constant temperature.

Current vs. Potential Difference

The resistance of the conductor is

(1) 1.0 Ω   (3) 0.50 Ω
(2) 2.0 Ω   (4) 4.0 Ω

Appendix B: 2010 Regents Exam

42 A student throws a baseball vertically upward and then catches it. If vertically upward is considered to be the positive direction, which graph best represents the relationship between velocity and time for the baseball? [Neglect friction.]

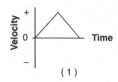

(1)

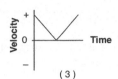

(3)

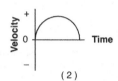

(2)

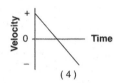

(4)

43 A 5.0-kilogram sphere, starting from rest, falls freely 22 meters in 3.0 seconds near the surface of a planet. Compared to the acceleration due to gravity near Earth's surface, the acceleration due to gravity near the surface of the planet is approximately

(1) the same   (3) one-half as great
(2) twice as great   (4) four times as great

44 A 15.0-kilogram mass is moving at 7.50 meters per second on a horizontal, frictionless surface. What is the total work that must be done on the mass to increase its speed to 11.5 meters per second?

(1) 120. J   (3) 570. J
(2) 422 J   (4) 992 J

45 The circuit diagram below represents four resistors connected to a 12-volt source.

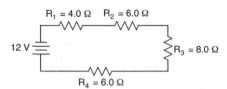

What is the total current in the circuit?

(1) 0.50 A   (3) 8.6 A
(2) 2.0 A   (4) 24 A

46 Which graph best represents the relationship between the power expended by a resistor that obeys Ohm's Law and the potential difference applied to the resistor?

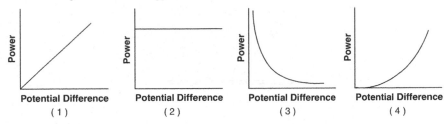

Potential Difference
(1)

Potential Difference
(2)

Potential Difference
(3)

Potential Difference
(4)

47 The distance between an electron and a proton is varied. Which pair of graphs best represents the relationship between gravitational force, F_g, and distance, r, and the relationship between electrostatic force, F_e, and distance, r, for these particles?

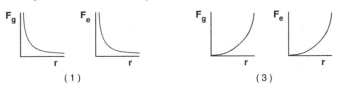

(1)   (3)

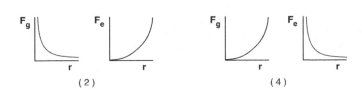

(2)   (4)

48 The diagram below represents a periodic wave traveling through a uniform medium.

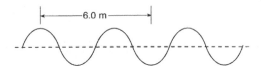

If the frequency of the wave is 2.0 hertz, the speed of the wave is

(1) 6.0 m/s   (3) 8.0 m/s
(2) 2.0 m/s   (4) 4.0 m/s

Appendix B: 2010 Regents Exam

49 The diagram below represents a light ray reflecting from a plane mirror.

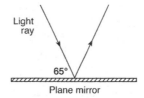

Light
ray

65°

Plane mirror

The angle of reflection for the light ray is

(1) 25°

(2) 35°

(3) 50.°

(4) 65°

50 The diagram below shows a standing wave in a string clamped at each end.

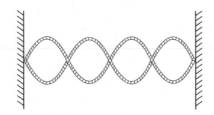

What is the total number of nodes and antinodes in the standing wave?

(1) 3 nodes and 2 antinodes

(2) 2 nodes and 3 antinodes

(3) 5 nodes and 4 antinodes

(4) 4 nodes and 5 antinodes

Appendix B: 2010 Regents Exam

Answer all questions in this part.

Directions (51–60): Record your answers in the spaces provided in your answer booklet.

Base your answers to questions 51 through 53 on the information and graph below.

A machine fired several projectiles at the same angle, θ, above the horizontal. Each projectile was fired with a different initial velocity, v_i. The graph below represents the relationship between the magnitude of the initial vertical velocity, v_{iy}, and the magnitude of the corresponding initial velocity, v_i, of these projectiles.

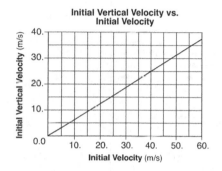

Initial Vertical Velocity vs. Initial Velocity

51 Determine the magnitude of the initial vertical velocity of the projectile, v_{iy}, when the magnitude of its initial velocity, v_i, was 40. meters per second. [1]

52 Determine the angle, θ, above the horizontal at which the projectiles were fired. [1]

53 Calculate the magnitude of the initial horizontal velocity of the projectile, v_{ix}, when the magnitude of its initial velocity, v_i, was 40. meters per second. [Show all work, including the equation and substitution with units.] [2]

54 A student makes a simple pendulum by attaching a mass to the free end of a 1.50-meter length of string suspended from the ceiling of her physics classroom. She pulls the mass up to her chin and releases it from rest, allowing the pendulum to swing in its curved path. Her classmates are surprised that the mass doesn't reach her chin on the return swing, even though she does not move. Explain why the mass does *not* have enough energy to return to its starting position and hit the girl on the chin. [1]

Appendix B: 2010 Regents Exam

55 A 6-ohm resistor and a 4-ohm resistor are connected in series with a 6-volt battery in an operating electric circuit. A voltmeter is connected to measure the potential difference across the 6-ohm resistor.
In the space *in your answer booklet*, draw a diagram of this circuit including the battery, resistors, and voltmeter using symbols from the *Reference Tables for Physical Setting/Physics*. Label each resistor with its value. [Assume the availability of any number of wires of negligible resistance.] [2]

56 When a spring is compressed 2.50×10^{-2} meter from its equilibrium position, the total potential energy stored in the spring is 1.25×10^{-2} joule. Calculate the spring constant of the spring. [Show all work, including the equation and substitution with units.] [2]

Base your answers to questions 57 and 58 on the information below.

A 3.50-meter length of wire with a cross-sectional area of 3.14×10^{-6} meter^2 is at 20° Celsius. The current in the wire is 24.0 amperes when connected to a 1.50-volt source of potential difference.

57 Determine the resistance of the wire. [1]

58 Calculate the resistivity of the wire. [Show all work, including the equation and substitution with units.] [2]

Base your answers to questions 59 and 60 on the information below.

In an experiment, a 0.028-kilogram rubber stopper is attached to one end of a string. A student whirls the stopper overhead in a horizontal circle with a radius of 1.0 meter. The stopper completes 10. revolutions in 10. seconds.

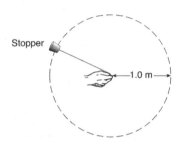

Stopper

1.0 m

(Not drawn to scale)

59 Determine the speed of the whirling stopper. [1]

60 Calculate the magnitude of the centripetal force on the whirling stopper. [Show all work, including the equation and substitution with units.] [2]

Appendix B: 2010 Regents Exam

Answer all questions in this part.

Directions (61–75): Record your answers in the spaces provided in your answer booklet.

Base your answers to questions 61 through 64 on the information below.

In a laboratory investigation, a student applied various downward forces to a vertical spring. The applied forces and the corresponding elongations of the spring from its equilibrium position are recorded in the data table below.

Data Table

Force (N)	Elongation (m)
0	0
0.5	0.010
1.0	0.018
1.5	0.027
2.0	0.035
2.5	0.046

Directions (61–63): Construct a graph on the grid *in your answer booklet*, following the directions below.

61 Mark an appropriate scale on the axis labeled "Force (N)." [1]

62 Plot the data points for force versus elongation. [1]

63 Draw the best-fit line or curve. [1]

64 Using your graph, calculate the spring constant of this spring. [Show all work, including the equation and substitution with units.] [2]

Base your answers to questions 65 through 68 on the information below.

An ice skater applies a horizontal force to a 20.-kilogram block on frictionless, level ice, causing the block to accelerate uniformly at 1.4 meters per second^2 to the right. After the skater stops pushing the block, it slides onto a region of ice that is covered with a thin layer of sand. The coefficient of kinetic friction between the block and the sand-covered ice is 0.28.

65 Calculate the magnitude of the force applied to the block by the skater. [Show all work, including the equation and substitution with units.] [2]

66 On the diagram *in your answer booklet*, starting at point A, draw a vector to represent the force applied to the block by the skater. Begin the vector at point A and use a scale of 1.0 centimeter = 5.0 newtons. [1]

67 Determine the magnitude of the normal force acting on the block. [1]

68 Calculate the magnitude of the force of friction acting on the block as it slides over the sand-covered ice. [Show all work, including the equation and substitution with units.] [2]

Appendix B: 2010 Regents Exam

Base your answers to questions 69 through 71 on the information and diagram below.

A monochromatic light ray ($f = 5.09 \times 10^{14}$ Hz) traveling in air is incident on the surface of a rectangular block of Lucite.

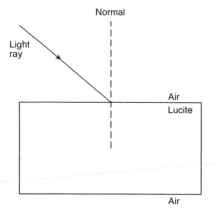

69 Measure the angle of incidence for the light ray to the *nearest degree*. [1]

70 Calculate the angle of refraction of the light ray when it enters the Lucite block. [Show all work, including the equation and substitution with units.] [2]

71 What is the angle of refraction of the light ray as it emerges from the Lucite block back into air? [1]

Base your answers to questions 72 through 75 on the information below.

As a mercury atom absorbs a photon of energy, an electron in the atom changes from energy level d to energy level e.

72 Determine the energy of the absorbed photon in electronvolts. [1]

73 Express the energy of the absorbed photon in joules. [1]

74 Calculate the frequency of the absorbed photon. [Show all work, including the equation and substitution with units.] [2]

75 Based on your calculated value of the frequency of the absorbed photon, determine its classification in the electromagnetic spectrum. [1]

Appendix B: 2010 Regents Exam

	Part A						Part B	
1)	2	16)	3	31)	3		36)	2
2)	3	17)	2	32)	4		37)	4
3)	3	18)	2	33)	1		38)	2
4)	4	19)	1	34)	3		39)	4
5)	4	20)	2	35)	2		40)	3
6)	1	21)	3				41)	2
7)	2	22)	2				42)	4
8)	1	23)	3				43)	3
9)	4	24)	4				44)	3
10)	4	25)	2				45)	1
11)	3	26)	1				46)	4
12)	3	27)	4				47)	1
13)	2	28)	3				48)	3
14)	4	29)	1				49)	1
15)	3	30)	1				50)	3

51 [1] Allow 1 credit for 25 m/s ± 1 m/s.

52 [1] Allow 1 credit for 39° ± 2°.

> **Note:** Allow credit for an answer that is consistent with the student's response to question 51.

53 [2] Allow a maximum of 2 credits. Refer to *Scoring Criteria for Calculations* in this rating guide.

Examples of 2-credit responses:

$$v_{ix} = v_i \cos \theta$$
$$v_{ix} = (40. \text{ m/s}) \cos 39°$$
$$v_{ix} = 31 \text{ m/s}$$

or

$$v_{ix}^2 + v_{iy}^2 = v_i^2$$
$$v_{ix} = \sqrt{v_i^2 - v_{iy}^2}$$
$$v_{ix} = \sqrt{(40. \text{ m/s})^2 - (25 \text{ m/s})^2}$$
$$v_{ix} = 31 \text{ m/s}$$

> **Note:** Allow credit for an answer that is consistent with the student's response to question 51 or 52

Appendix B: 2010 Regents Exam

54 [1] Allow 1 credit. Acceptable responses include, but are not limited to:

— friction

— Some of the gravitational energy of the mass was converted into internal energy. Therefore, it could not return to its original height.

— air resistance

55 [2] Allow a maximum of 2 credits, allocated as follows:

• Allow 1 credit for drawing a series circuit containing two resistors and a battery.

• Allow 1 credit for correct placement of the voltmeter.

Example of a 2-credit response:

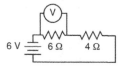

Note: Allow credit even if the student draws a cell instead of a battery and/or labels only one resistor with its value.

56 [2] Allow a maximum of 2 credits. Refer to *Scoring Criteria for Calculations* in this rating guide.

Example of a 2-credit response:

$$PE_s = \frac{1}{2}kx^2$$

$$k = \frac{2PE_s}{x^2}$$

$$k = \frac{2(1.25 \times 10^{-2} \text{ J})}{(2.50 \times 10^{-2} \text{ m})^2}$$

$$k = 40.0 \text{ N/m}$$

57 [1] Allow 1 credit for $6.25 \times 10^{-2}\ \Omega$.

58 [2] Allow a maximum of 2 credits. Refer to *Scoring Criteria for Calculations* in this rating guide.

Example of a 2-credit response:

$$R = \frac{\rho L}{A}$$

$$\rho = \frac{RA}{L}$$

$$\rho = \frac{(6.25 \times 10^{-2}\ \Omega)(3.14 \times 10^{-6}\ \text{m}^2)}{3.50\,\text{m}}$$

$$\rho = 5.61 \times 10^{-8}\ \Omega \bullet \text{m}$$

Note: Allow credit for an answer that is consistent with the student's response to question 57.

59 [1] Allow 1 credit for 6.3 m/s.

Appendix B: 2010 Regents Exam

60 [2] Allow a maximum of 2 credits. Refer to *Scoring Criteria for Calculations* in this rating guide.

Example of a 2-credit response:

$$F_c = ma_c \qquad a_c = \frac{v^2}{r}$$

$$F_c = \frac{mv^2}{r}$$

$$F_c = \frac{(0.028 \text{ kg})(6.3 \text{ m/s})^2}{1.0 \text{ m}}$$

$$F_c = 1.1 \text{ N}$$

Note: Allow credit for an answer that is consistent with the student's response to question 59.

61 [1] Allow 1 credit for an appropriate linear scale.

62 [1] Allow 1 credit for plotting all points accurately ± 0.3 grid space.

63 [1] Allow 1 credit for drawing the best-fit line or curve consistent with the student's responses to questions 61 and 62.

Example of a 3-credit graph for questions 61–63:

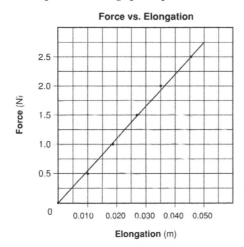

Force vs. Elongation

64 [2] Allow a maximum of 2 credits. Refer to *Scoring Criteria for Calculations* in this rating guide.

Examples of 2-credit responses:

$$k = \frac{\Delta F}{\Delta x}$$

$$k = \frac{2.5 \text{ N}}{0.046 \text{ m}}$$

$$k = 54 \text{ N/m}$$

or

$$\text{slope} = \frac{\Delta y}{\Delta x}$$

$$\text{slope} = \frac{2.5 \text{ N} - 0.8 \text{ N}}{0.046 \text{ m} - 0.015 \text{ m}}$$

$$\text{slope} = 55 \text{ N/m}$$

Note: Allow credit for an answer that is consistent with the student's graph.
The slope may be determined by substitution of values from the data table only if the data points are on the best-fit line *or* if the student failed to draw a best-fit line.

Appendix B: 2010 Regents Exam

65 [2] Allow a maximum of 2 credits. Refer to *Scoring Criteria for Calculations* in this rating guide.

Example of a 2-credit response:

$$a = \frac{F_{net}}{m}$$
$$F_{net} = ma$$
$$F_{net} = (20.\ \text{kg})(1.4\ \text{m/s}^2)$$
$$F_{net} = 28\ \text{N}$$

66 [1] Allow 1 credit for a vector 5.6 cm ± 0.2 cm long parallel to the surface of the ice and pointing to the right.

Example of a 1-credit response:

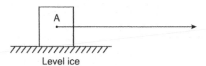

Level ice

Note: Allow credit for an answer that is consistent with the student's response to question 65. The vector need *not* start at point A to receive this credit.

67 [1] Allow 1 credit for 2.0×10^2 N *or* 196 N.

68 [2] Allow a maximum of 2 credits. Refer to *Scoring Criteria for Calculations* in this rating guide.

Example of a 2-credit response:

$$F_f = \mu F_N$$
$$F_f = (0.28)(2.0 \times 10^2\ \text{N})$$
$$F_f = 56\ \text{N}$$

Note: Allow credit for an answer that is consistent with the student's response to question 67.

Appendix B: 2010 Regents Exam

69 [1] Allow 1 credit for 50.° ± 2°.

70 [2] Allow a maximum of 2 credits. Refer to *Scoring Criteria for Calculations* in this rating guide.

Example of a 2-credit response:

$n_1 \sin \theta_1 = n_2 \sin \theta_2$

$\sin \theta_2 = \dfrac{n_1 \sin \theta_1}{n_2}$

$\sin \theta_2 = \dfrac{1.00 (\sin 50.°)}{1.50}$

$\theta_2 = 31°$

Note: Allow credit for an answer that is consistent with the student's response to question 69.

71 [1] Allow 1 credit for 50.°.

Note: Allow credit for an answer that is consistent with the student's response to question 69 or 70.

72 [1] Allow 1 credit for 1.24 eV.

73 [1] Allow 1 credit for 1.98×10^{-19} J *or* an answer that is consistent with the student's response to question 72.

74 [2] Allow a maximum of 2 credits. Refer to *Scoring Criteria for Calculations* in this rating guide.

Example of a 2-credit response:

$E_{photon} = hf$

$f = \dfrac{E_{photon}}{h}$

$f = \dfrac{1.98 \times 10^{-19} \text{ J}}{6.63 \times 10^{-34} \text{ J} \cdot \text{s}}$

$f = 2.99 \times 10^{14}$ Hz

Note: Allow credit for an answer that is consistent with the student's response to question 73.

75 [1] Allow 1 credit for infrared *or* an answer that is consistent with the student's response to question 74.

Appendix B: 2010 Regents Exam

Appendix C: 2009 Regents Exam

Directions (1–35): For *each* statement or question, write on the separate answer sheet the *number* of the word or expression that, of those given, best completes the statement or answers the question.

1 On a highway, a car is driven 80. kilometers during the first 1.00 hour of travel, 50. kilometers during the next 0.50 hour, and 40. kilometers in the final 0.50 hour. What is the car's average speed for the entire trip?

 (1) 45 km/h   (3) 85 km/h
 (2) 60. km/h   (4) 170 km/h

2 The vector diagram below represents the horizontal component, F_H, and the vertical component, F_V, of a 24-newton force acting at 35° above the horizontal.

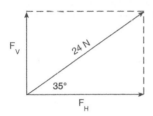

What are the magnitudes of the horizontal and vertical components?

 (1) F_H = 3.5 N and F_V = 4.9 N
 (2) F_H = 4.9 N and F_V = 3.5 N
 (3) F_H = 14 N and F_V = 20. N
 (4) F_H = 20. N and F_V = 14 N

3 Which quantity is a vector?

 (1) impulse   (3) speed
 (2) power   (4) time

4 A high-speed train in Japan travels a distance of 300. kilometers in 3.60×10^3 seconds. What is the average speed of this train?

 (1) 1.20×10^{-2} m/s   (3) 12.0 m/s
 (2) 8.33×10^{-2} m/s   (4) 83.3 m/s

5 A 25-newton weight falls freely from rest from the roof of a building. What is the total distance the weight falls in the first 1.0 second?

 (1) 19.6 m   (3) 4.9 m
 (2) 9.8 m   (4) 2.5 m

6 A golf ball is given an initial speed of 20. meters per second and returns to level ground. Which launch angle above level ground results in the ball traveling the greatest horizontal distance? [Neglect friction.]

 (1) 60.°   (3) 30.°
 (2) 45°   (4) 15°

Base your answers to questions 7 and 8 on the information below.

A go-cart travels around a flat, horizontal, circular track with a radius of 25 meters. The mass of the go-cart with the rider is 200. kilograms. The magnitude of the maximum centripetal force exerted by the track on the go-cart is 1200. newtons.

7 What is the maximum speed the 200.-kilogram go-cart can travel without sliding off the track?

 (1) 8.0 m/s   (3) 150 m/s
 (2) 12 m/s   (4) 170 m/s

8 Which change would increase the maximum speed at which the go-cart could travel without sliding off this track?

 (1) Decrease the coefficient of friction between the go-cart and the track.
 (2) Decrease the radius of the track.
 (3) Increase the radius of the track.
 (4) Increase the mass of the go-cart.

Appendix C: 2009 Regents Exam

9 A 0.50-kilogram cart is rolling at a speed of 0.40 meter per second. If the speed of the cart is doubled, the inertia of the cart is

(1) halved   (3) quadrupled
(2) doubled   (4) unchanged

10 Two forces, F_1 and F_2, are applied to a block on a frictionless, horizontal surface as shown below.

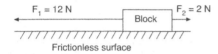

Frictionless surface

If the magnitude of the block's acceleration is 2.0 meters per second², what is the mass of the block?

(1) 1 kg   (3) 6 kg
(2) 5 kg   (4) 7 kg

11 Which body is in equilibrium?

(1) a satellite orbiting Earth in a circular orbit
(2) a ball falling freely toward the surface of Earth
(3) a car moving with a constant speed along a straight, level road
(4) a projectile at the highest point in its trajectory

12 What is the weight of a 2.00-kilogram object on the surface of Earth?

(1) 4.91 N   (3) 9.81 N
(2) 2.00 N   (4) 19.6 N

13 A 70.-kilogram cyclist develops 210 watts of power while pedaling at a constant velocity of 7.0 meters per second east. What average force is exerted eastward on the bicycle to maintain this constant speed?

(1) 490 N   (3) 3.0 N
(2) 30. N   (4) 0 N

14 The gravitational potential energy, with respect to Earth, that is possessed by an object is dependent on the object's

(1) acceleration   (3) position
(2) momentum   (4) speed

Note that question 15 has only three choices.

15 As a ball falls freely toward the ground, its total mechanical energy

(1) decreases
(2) increases
(3) remains the same

16 A spring with a spring constant of 4.0 newtons per meter is compressed by a force of 1.2 newtons. What is the total elastic potential energy stored in this compressed spring?

(1) 0.18 J   (3) 0.60 J
(2) 0.36 J   (4) 4.8 J

17 A distance of 1.0 meter separates the centers of two small charged spheres. The spheres exert gravitational force F_g and electrostatic force F_e on each other. If the distance between the spheres' centers is increased to 3.0 meters, the gravitational force and electrostatic force, respectively, may be represented as

(1) $\dfrac{F_g}{9}$ and $\dfrac{F_e}{9}$ (3) $3F_g$ and $3F_e$

(2) $\dfrac{F_g}{3}$ and $\dfrac{F_e}{3}$ (4) $9F_g$ and $9F_e$

18 The electrical resistance of a metallic conductor is inversely proportional to its

(1) temperature   (3) cross-sectional area
(2) length   (4) resistivity

19 In a simple electric circuit, a 24-ohm resistor is connected across a 6.0-volt battery. What is the current in the circuit?

(1) 1.0 A   (3) 140 A
(2) 0.25 A   (4) 4.0 A

20 An operating 100.-watt lamp is connected to a 120-volt outlet. What is the total electrical energy used by the lamp in 60. seconds?

(1) 0.60 J   (3) 6.0×10^3 J
(2) 1.7 J   (4) 7.2×10^3 J

21 A beam of electrons is directed into the electric field between two oppositely charged parallel plates, as shown in the diagram below.

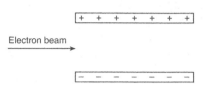

Electron beam

The electrostatic force exerted on the electrons by the electric field is directed

(1) into the page
(2) out of the page
(3) toward the bottom of the page
(4) toward the top of the page

22 When two ring magnets are placed on a pencil, magnet A remains suspended above magnet B, as shown below.

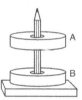

Which statement describes the gravitational force and the magnetic force acting on magnet A due to magnet B?

(1) The gravitational force is attractive and the magnetic force is repulsive.
(2) The gravitational force is repulsive and the magnetic force is attractive.
(3) Both the gravitational force and the magnetic force are attractive.
(4) Both the gravitational force and the magnetic force are repulsive.

23 Which color of light has a wavelength of 5.0×10^{-7} meter in air?

(1) blue   (3) orange
(2) green   (4) violet

24 Which type of wave requires a material medium through which to travel?

(1) sound   (3) television
(2) radio   (4) x ray

25 A periodic wave is produced by a vibrating tuning fork. The amplitude of the wave would be greater if the tuning fork were

(1) struck more softly
(2) struck harder
(3) replaced by a lower frequency tuning fork
(4) replaced by a higher frequency tuning fork

26 The sound wave produced by a trumpet has a frequency of 440 hertz. What is the distance between successive compressions in this sound wave as it travels through air at STP?

(1) 1.5×10^{-6} m   (3) 1.3 m
(2) 0.75 m   (4) 6.8×10^{5} m

27 The diagram below represents a light ray striking the boundary between air and glass.

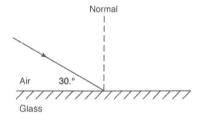

What would be the angle between this light ray and its reflected ray?

(1) 30.°   (3) 120.°
(2) 60.°   (4) 150.°

28 In which way does blue light change as it travels from diamond into crown glass?

(1) Its frequency decreases.
(2) Its frequency increases.
(3) Its speed decreases.
(4) Its speed increases.

29 The diagram below shows two pulses approaching each other in a uniform medium.

Which diagram best represents the superposition of the two pulses?

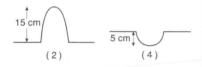

30 Sound waves strike a glass and cause it to shatter. This phenomenon illustrates

(1) resonance   (3) reflection
(2) refraction   (4) diffraction

31 An alpha particle consists of two protons and two neutrons. What is the charge of an alpha particle?

(1) 1.25×10^{19} C   (3) 6.40×10^{-19} C
(2) 2.00 C   (4) 3.20×10^{-19} C

32 An electron in the c level of a mercury atom returns to the ground state. Which photon energy could *not* be emitted by the atom during this process?

(1) 0.22 eV   (3) 4.86 eV
(2) 4.64 eV   (4) 5.43 eV

33 Which phenomenon provides evidence that light has a wave nature?

(1) emission of light from an energy-level transition in a hydrogen atom
(2) diffraction of light passing through a narrow opening
(3) absorption of light by a black sheet of paper
(4) reflection of light from a mirror

34 When Earth and the Moon are separated by a distance of 3.84×10^8 meters, the magnitude of the gravitational force of attraction between them is 2.0×10^{20} newtons. What would be the magnitude of this gravitational force of attraction if Earth and the Moon were separated by a distance of 1.92×10^8 meters?

(1) 5.0×10^{19} N   (3) 4.0×10^{20} N
(2) 2.0×10^{20} N   (4) 8.0×10^{20} N

35 The particles in a nucleus are held together primarily by the

(1) strong force   (3) electrostatic force
(2) gravitational force   (4) magnetic force

Appendix C: 2009 Regents Exam

Answer all questions in this part.

Directions (36–47): For *each* statement or question, write on the separate answer sheet the *number* of the word or expression that, of those given, best completes the statement or answers the question.

36 The work done in lifting an apple one meter near Earth's surface is approximately

(1) 1 J   (3) 100 J
(2) 0.01 J   (4) 1000 J

Base your answers to questions 37 and 38 on the graph below, which represents the motion of a car during a 6.0-second time interval.

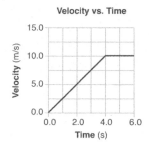

Velocity vs. Time

37 What is the acceleration of the car at $t = 5.0$ seconds?

(1) 0.0 m/s^2 (3) 2.5 m/s^2
(2) 2.0 m/s^2 (4) 10. m/s^2

38 What is the total distance traveled by the car during this 6.0-second interval?

(1) 10. m   (3) 40. m
(2) 20. m   (4) 60. m

39 A person weighing 785 newtons on the surface of Earth would weigh 298 newtons on the surface of Mars. What is the magnitude of the gravitational field strength on the surface of Mars?

(1) 2.63 N/kg   (3) 6.09 N/kg
(2) 3.72 N/kg   (4) 9.81 N/kg

40 A motorcycle being driven on a dirt path hits a rock. Its 60.-kilogram cyclist is projected over the handlebars at 20. meters per second into a haystack. If the cyclist is brought to rest in 0.50 second, the magnitude of the average force exerted on the cyclist by the haystack is

(1) 6.0×10^1 N   (3) 1.2×10^3 N
(2) 5.9×10^2 N   (4) 2.4×10^3 N

Base your answers to questions 41 and 42 on the information below.

A boy pushes his wagon at constant speed along a level sidewalk. The graph below represents the relationship between the horizontal force exerted by the boy and the distance the wagon moves.

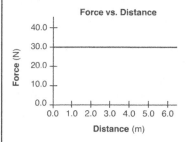

Force vs. Distance

41 What is the total work done by the boy in pushing the wagon 4.0 meters?

(1) 5.0 J   (3) 120 J
(2) 7.5 J   (4) 180 J

42 As the boy pushes the wagon, what happens to the wagon's energy?

(1) Gravitational potential energy increases.
(2) Gravitational potential energy decreases.
(3) Internal energy increases.
(4) Internal energy decreases.

43 Which is an SI unit for work done on an object?

(1) $\dfrac{\text{kg} \cdot \text{m}^2}{\text{s}^2}$

(3) $\dfrac{\text{kg} \cdot \text{m}}{\text{s}}$

(2) $\dfrac{\text{kg} \cdot \text{m}^2}{\text{s}}$

(4) $\dfrac{\text{kg} \cdot \text{m}}{\text{s}^2}$

44 The momentum of a photon, p, is given by the equation $p = \dfrac{h}{\lambda}$ where h is Planck's constant and λ is the photon's wavelength. Which equation correctly represents the energy of a photon in terms of its momentum?

(1) $E_{photon} = phc$

(3) $E_{photon} = \dfrac{p}{c}$

(2) $E_{photon} = \dfrac{hp}{c}$

(4) $E_{photon} = pc$

45 A constant potential difference is applied across a variable resistor held at constant temperature. Which graph best represents the relationship between the resistance of the variable resistor and the current through it?

(1)

(3)

(2)

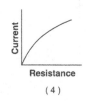

(4)

46 A 3.0-ohm resistor and a 6.0-ohm resistor are connected in series in an operating electric circuit. If the current through the 3.0-ohm resistor is 4.0 amperes, what is the potential difference across the 6.0-ohm resistor?

(1) 8.0 V   (3) 12 V
(2) 2.0 V   (4) 24 V

47 Which combination of resistors has the *smallest* equivalent resistance?

(1)

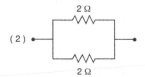

(2)

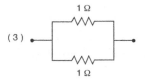

(3)

(4)

Appendix C: 2009 Regents Exam

Part B–2

Answer all questions in this part.

Directions (48–59): Record your answers in the spaces provided in your answer booklet.

48 A cart travels 4.00 meters east and then 4.00 meters north. Determine the magnitude of the cart's resultant displacement. [1]

49 A 70-kilogram hockey player skating east on an ice rink is hit by a 0.1-kilogram hockey puck moving toward the west. The puck exerts a 50-newton force toward the west on the player. Determine the magnitude of the force that the player exerts on the puck during this collision. [1]

50 On a snow-covered road, a car with a mass of 1.1×10^3 kilograms collides head-on with a van having a mass of 2.5×10^3 kilograms traveling at 8.0 meters per second. As a result of the collision, the vehicles lock together and immediately come to rest. Calculate the speed of the car immediately before the collision. [Neglect friction.] [Show all work, including the equation and substitution with units.] [2]

51 A baby and stroller have a total mass of 20. kilograms. A force of 36 newtons keeps the stroller moving in a circular path with a radius of 5.0 meters. Calculate the speed at which the stroller moves around the curve. [Show all work, including the equation and substitution with units.] [2]

52 A 10.-newton force compresses a spring 0.25 meter from its equilibrium position. Calculate the spring constant of this spring. [Show all work, including the equation and substitution with units.] [2]

53 Two oppositely charged parallel metal plates, 1.00 centimeter apart, exert a force with a magnitude of 3.60×10^{-15} newton on an electron placed between the plates. Calculate the magnitude of the electric field strength between the plates. [Show all work, including the equation and substitution with units.] [2]

54 On the diagram *in your answer booklet*, sketch *at least four* electric field lines with arrowheads that represent the electric field around a negatively charged conducting sphere. [1]

55 In the space *in your answer booklet*, draw a diagram of an operating circuit that includes:
- a battery as a source of potential difference
- *two* resistors in parallel with each other
- an ammeter that reads the total current in the circuit [2]

56 Calculate the resistance of a 900.-watt toaster operating at 120 volts. [Show all work, including the equation and substitution with units.] [2]

57 A student and a physics teacher hold opposite ends of a horizontal spring stretched from west to east along a tabletop. Identify the directions in which the student should vibrate the end of the spring to produce transverse periodic waves. [1]

Appendix C: 2009 Regents Exam

Base your answers to questions 58 and 59 on the information and diagram below.

The vertical lines in the diagram represent compressions in a sound wave of constant frequency propagating to the right from a speaker toward an observer at point A.

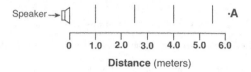

58 Determine the wavelength of this sound wave. [1]

59 The speaker is then moved at constant speed toward the observer at A. Compare the wavelength of the sound wave received by the observer while the speaker is moving to the wavelength observed when the speaker was at rest. [1]

Part C

Answer all questions in this part.

Directions (60–72): Record your answers in the spaces provided in your answer booklet.

Base your answers to questions 60 through 62 on the information below.

The path of a stunt car driven horizontally off a cliff is represented in the diagram below. After leaving the cliff, the car falls freely to point A in 0.50 second and to point B in 1.00 second.

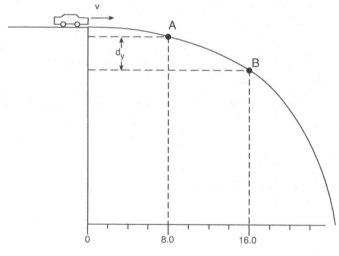

Distance From Base of Cliff (m)

60 Determine the magnitude of the horizontal component of the velocity of the car at point B. [Neglect friction.] [1]

61 Determine the magnitude of the vertical velocity of the car at point A. [1]

62 Calculate the magnitude of the vertical displacement, d_y, of the car from point A to point B. [Neglect friction.] [Show all work, including the equation and substitution with units.] [2]

Appendix C: 2009 Regents Exam

Base your answers to questions 63 through 65 on the information below.

A roller coaster car has a mass of 290. kilograms. Starting from rest, the car acquires 3.13×10^5 joules of kinetic energy as it descends to the bottom of a hill in 5.3 seconds.

63 Calculate the height of the hill. [Neglect friction.] [Show all work, including the equation and substitution with units.] [2]

64 Calculate the speed of the roller coaster car at the bottom of the hill. [Show all work, including the equation and substitution with units.] [2]

65 Calculate the magnitude of the average acceleration of the roller coaster car as it descends to the bottom of the hill. [Show all work, including the equation and substitution with units.] [2]

Base your answers to questions 66 and 67 on the information below.

One end of a rope is attached to a variable speed drill and the other end is attached to a 5.0-kilogram mass. The rope is draped over a hook on a wall opposite the drill. When the drill rotates at a frequency of 20.0 Hz, standing waves of the same frequency are set up in the rope. The diagram below shows such a wave pattern.

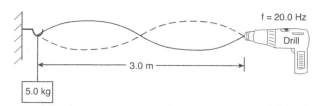

66 Determine the wavelength of the waves producing the standing wave pattern. [1]

67 Calculate the speed of the wave in the rope. [Show all work, including the equation and substitution with units.] [2]

Appendix C: 2009 Regents Exam

Base your answers to questions 68 and 69 on the information below.

A ray of monochromatic light (f = 5.09 × 10^{14} Hz) passes from air into Lucite at an angle of incidence of 30.°.

68 Calculate the angle of refraction in the Lucite. [Show all work, including the equation and substitution with units.] [2]

69 Using a protractor and straightedge, on the diagram *in your answer booklet*, draw the refracted ray in the Lucite. [1]

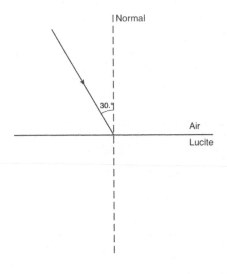

Base your answers to questions 70 through 72 on the information below.

A photon with a frequency of 5.48 × 10^{14} hertz is emitted when an electron in a mercury atom falls to a lower energy level.

70 Identify the color of light associated with this photon. [1]

71 Calculate the energy of this photon in joules. [Show all work, including the equation and substitution with units.] [2]

72 Determine the energy of this photon in electronvolts. [1]

Appendix C: 2009 Regents Exam

	Part A						Part B	
1)	3	13)	2	25)	2	36)	1
2)	4	14)	3	26)	2	37)	1
3)	1	15)	3	27)	3	38)	3
4)	4	16)	1	28)	4	39)	2
5)	3	17)	1	29)	2	40)	4
6)	2	18)	3	30)	1	41)	3
7)	2	19)	2	31)	4	42)	3
8)	3	20)	3	32)	4	43)	1
9)	4	21)	4	33)	2	44)	4
10)	2	22)	1	34)	4	45)	1
11)	3	23)	2	35)	1	46)	4
12)	4	24)	1			47)	3

Part B–2

48 [1] Allow 1 credit for 5.66 m.

49 [1] Allow 1 credit for 50 N.

50 [2] Allow a maximum of 2 credits. Refer to *Scoring Criteria for Calculations* in this rating guide.

Examples of 2-credit responses:

$$p_{before} = p_{after}$$
$$m_1 v_1 + m_2 v_2 = 0$$
$$v_1 = \frac{-m_2 v_2}{m_1}$$
$$v_1 = \frac{-(2.5 \times 10^3 \text{ kg})(8.0 \text{ m/s })}{1.1 \times 10^3 \text{ kg}}$$
$$v_1 = -18 \text{ m/s} \quad or \quad 18 \text{ m/s}$$

or

$$m_1 v_1 = m_2 v_2$$
$$(1.1 \times 10^3 \text{ kg}) v_1 = (2.5 \times 10^3 \text{ kg})(8.0 \text{ m/s})$$
$$v_1 = 18 \text{ m/s}$$

Appendix C: 2009 Regents Exam

51 [2]   Allow a maximum of 2 credits. Refer to *Scoring Criteria for Calculations* in this rating guide.

Example of a 2-credit response:

$$F_c = \frac{mv^2}{r}$$

$$v = \sqrt{\frac{F_c r}{m}}$$

$$v = \sqrt{\frac{(36\,N)(5.0\,m)}{20.\,kg}}$$

$$v = 3.0\,m/s$$

52 [2]   Allow a maximum of 2 credits. Refer to *Scoring Criteria for Calculations* in this rating guide.

Example of a 2-credit response:

$$F_s = kx$$

$$k = \frac{F_s}{x}$$

$$k = \frac{10.\,N}{0.25\,m}$$

$$k = 40.\,N/m$$

53 [2]   Allow a maximum of 2 credits. Refer to *Scoring Criteria for Calculations* in this rating guide.

Example of a 2-credit response:

$$E = \frac{F_e}{q}$$

$$E = \frac{3.60 \times 10^{-15}\,N}{1.60 \times 10^{-19}\,C}$$

$$E = 2.25 \times 10^4\,N/C$$

Appendix C: 2009 Regents Exam

54 [1] Allow 1 credit for *at least four* straight lines drawn perpendicular to the surface of the sphere with each line having an arrowhead directed toward the sphere and ending within 0.2 cm of the sphere.

Example of a 1-credit response:

Note: Allow credit even if the lines are *not* drawn symmetrically.

55 [2] Allow a maximum of 2 credits, allocated as follows:

- Allow 1 credit for two resistors connected in parallel with the battery (or cell) in a complete circuit.
- Allow 1 credit for an ammeter connected in the circuit to measure the total current.

Example of a 2-credit response:

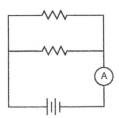

Examples of 1-credit responses:

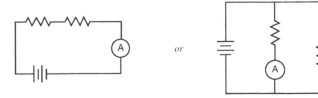

or

Note: Allow credit for lines *not* touching the battery if the distance from the lines to the battery is ≤ the distance between the battery symbol lines.

56 [2] Allow a maximum of 2 credits. Refer to *Scoring Criteria for Calculations* in this rating guide.

Example of a 2-credit response:

$$P = \frac{V^2}{R}$$

$$R = \frac{V^2}{P}$$

$$R = \frac{(120 \text{ V})^2}{900. \text{ W}}$$

$$R = 16 \ \Omega$$

57 [1] Allow 1 credit. Acceptable responses include, but are not limited to:

— north and south

— up and down

— perpendicular to spring

— left and right

Note: Do *not* allow credit for back and forth or east and west.

58 [1] Allow 1 credit for 1.5 m.

59 [1] Allow 1 credit for indicating that the wavelength is shorter while the speaker is moving *or* for an answer that is consistent with the student's response to question 58.

Appendix C: 2009 Regents Exam

Part C

60 [1] Allow 1 credit for 16 m/s.

61 [1] Allow 1 credit for 4.9 m/s.

62 [2] Allow a maximum of 2 credits. Refer to *Scoring Criteria for Calculations* in this rating guide.

Example of a 2-credit response:

$d = v_i t + \frac{1}{2} at^2$

$d_y = (4.9 \text{ m/s}) (0.50 \text{ s}) + \frac{1}{2}(9.81 \text{ m/s}^2) (0.50 \text{ s})^2$

$d_y = 3.7 \text{ m}$

Note: Allow credit for an answer that is consistent with the student's response to question 61.

63 [2] Allow a maximum of 2 credits. Refer to *Scoring Criteria for Calculations* in this rating guide.

Example of a 2-credit response:

$\Delta KE = \Delta PE = mg\Delta h$

$\Delta h = \dfrac{\Delta KE}{mg}$

$\Delta h = \dfrac{3.13 \times 10^5 \text{ J}}{(290. \text{ kg})(9.81 \text{ m/s}^2)}$

$\Delta h = 110. \text{ m}$

64 [2] Allow a maximum of 2 credits. Refer to *Scoring Criteria for Calculations* in this rating guide.

Example of a 2-credit response:

$$KE = \frac{1}{2} mv^2$$

$$v = \sqrt{\frac{2KE}{m}}$$

$$v = \sqrt{\frac{2(3.13 \times 10^5 \text{ J})}{290. \text{ kg}}}$$

$$v = 46.5 \text{ m/s}$$

65 [2] Allow a maximum of 2 credits. Refer to *Scoring Criteria for Calculations* in this rating guide.

Example of a 2-credit response:

$$a = \frac{\Delta v}{t}$$

$$a = \frac{46.5 \text{ m/s}}{5.3 \text{ s}}$$

$$a = 8.8 \text{ m/s}^2$$

Note: Allow credit for an answer that is consistent with the student's response to question 64.

66 [1] Allow 1 credit for 3.0 m *or* 3 m.

67 [2] Allow a maximum of 2 credits. Refer to *Scoring Criteria for Calculations* in this rating guide.

Example of a 2-credit response:

$$v = f\lambda$$

$$v = (20.0 \text{ Hz}) (3.0 \text{ m})$$

$$v = 60. \text{ m/s}$$

Note: Allow credit for an answer that is consistent with the student's response to question 66.

Appendix C: 2009 Regents Exam

68 [2] Allow a maximum of 2 credits. Refer to *Scoring Criteria for Calculations* in this rating guide.

Example of a 2-credit response:

$$n_1 \sin \theta_1 = n_2 \sin \theta_2$$
$$\sin \theta_2 = \frac{n_1 \sin \theta_1}{n_2}$$
$$\sin \theta_2 = \frac{(1.00)(\sin 30.°)}{1.50}$$
$$\theta_2 = 19°$$

69 [1] Allow 1 credit for a response correctly showing the refracted ray at 19° ± 2° to the normal.

Example of a 1 credit response:

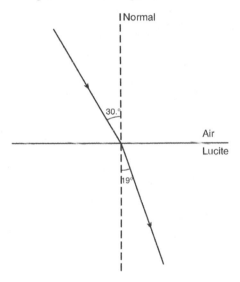

Note: Allow credit even if the arrowhead is missing.
Allow credit for an answer that is consistent with the student's response to question 68.

Appendix C: 2009 Regents Exam

70 [1] Allow 1 credit for green.

71 [2] Allow a maximum of 2 credits. Refer to *Scoring Criteria for Calculations* in the rating guide.

Example of a 2-credit response:

$E_{photon} = hf$

$E_{photon} = (6.63 \times 10^{-34} \text{ J} \bullet \text{s})(5.48 \times 10^{14} \text{ Hz})$

$E_{photon} = 3.63 \times 10^{-19} \text{ J}$

72 [1] Allow 1 credit for 2.27 eV.

Note: Allow credit for an answer that is consistent with the student's response to question 71.

Index

Index

Index

Index